Biodiversity
Threats and Outlook

Biodiversity Threats and Outlook

Dr. Kalu Ram

RANDOM PUBLICATIONS
NEW DELHI (INDIA)

Biodiversity Threats and Outlook

ISBN 978-93-5111-526-7

Published in 2015 in India by

RANDOM PUBLICATIONS

4376-A/4B, Gali Murari Lal, Ansari Road
New Delhi-110 002
Phone : +9111-43580356, 011-23289044, 011-43142548
e-mail: sales@randompublications.com,
info@randompublications.com, randomexports@gmail.com
Reprinted 2019

Type Setting by : Friends Media, Delhi-110089
Digitally Printed at : Replika Press Pvt. Ltd.

Preface

Biological diversity, or biodiversity reflects the number, variety and variability of living organisms and how these change from one location to another and over time. It includes diversity within species, between species and between ecosystems. Ecosystems, whose functioning depends on biodiversity, provide the basic necessities of life, offer protection from natural disasters and disease, and shape human cultures and spiritual beliefs. Besides those provisioning, regulating and cultural services they provide, ecosystems also support and maintain life processes such as biomass production and nutrient cycling which are essential to human well-being.

The impact of humans on the natural environment is significant and growing, causing changes in biodiversity that have been more rapid in the past 50 years than at any time before. As demographic pressures and consumption levels increase, biodiversity decreases, and the ability of the natural world to continue delivering the goods and services on which humanity ultimately depends may be undermined. Biodiversity loss disrupts the functioning of ecosystems, making them more vulnerable to perturbations and less able to supply humans with needed services. The consequences are often harshest on the rural poor, who depend most immediately upon ecosystem services for their livelihoods, and biodiversity loss poses a significant barrier to meeting the Millennium Development Goals.

This book provides a broad view of the many threats to global biodiversity imposed by human-induced changes and an overview of the policy responses required to combat them. It offers a solid description of the current understanding of threats to biodiversity with a range of illustrative examples—a valuable point of reference for ecologists, environmental scientists, and students as well as, policy-makers and all other environmental professionals.

Author

Contents

1

What is Biodiversity?

The degree of variation of life forms within a given species, ecosystem, biome, or an entire planet is called as biodiversity. Biodiversity is a measure of the health of ecosystems. Biodiversity is in part a function of climate. In terrestrial habitats, tropical regions are typically rich whereas polar regions support fewer species.

Rapid environmental changes typically cause mass extinctions. One estimate is that less than 1% of the species that have existed on Earth are extant. Since life began on Earth, five major mass extinctions and several minor events have led to large and sudden drops in biodiversity. The Phanerozoic eon marked a rapid growth in biodiversity via the Cambrian explosion—a period during which the majority of multicellular phyla first appeared. The next 400 million years included repeated, massive biodiversity losses classified as mass extinction events. In the Carboniferous, rainforest collapse led to a great loss of plant and animal life. The Permian–Triassic extinction event, 251 million years ago, was the worst; vertebrate recovery took 30 million years. The most recent, the Cretaceous–Paleogene extinction event, occurred 65 million years ago, and has often attracted more attention than others because it resulted in the extinction of the dinosaurs.

The period since the emergence of humans has displayed an ongoing biodiversity reduction and an accompanying loss of genetic diversity. Named the Holocene extinction, the reduction is caused primarily by human impacts, particularly habitat destruction. Conversely, biodiversity impacts human health in a number of ways, both positively and negatively.

Levels of Biodiversity

"Biodiversity" is most commonly used to replace the more clearly defined and long established terms, species diversity and species richness. Biologists most often define biodiversity as the "totality of genes, species, and ecosystems of a region". An advantage of this definition is that it seems to describe most circumstances and presents a unified view of the traditional three levels at which biological variety has been identified:

1. Species diversity
2. Ecosystem diversity
3. Genetic diversity

This multilevel construct is consistent with Dasmann and Lovejoy. An explicit definition consistent with this interpretation was first given in a paper by Bruce A. Wilcox commissioned by the International Union for the Conservation of Nature and Natural Resources (IUCN) for the 1982 World National Parks Conference. Wilcox's definition was "Biological diversity is the variety of life forms...at all levels of biological systems (i.e., molecular, organismic, population, species and ecosystem)...". The 1992 United Nations Earth Summit defined "biological diversity" as "the variability among living organisms from all sources, including, 'inter alia', terrestrial, marine, and other aquatic ecosystems, and the ecological complexes of which they are part: this includes diversity within species, between species and of ecosystems". This definition is used in the United Nations Convention on Biological Diversity.

Genetic Diversity

Genetic diversity is the sum of genetic information contained in the genes of individuals of plants, animals and micro-organisms. Each species is the repository of an immense amount of genetic information. The number of genes range from about 1000 in bacteria, up to 400 000 or more in many flowering plants.

Each species is made up of many organisms, and virtually no two members of the same species are genetically identical. This means for example that even if an endangered species is saved from extinction, it will probably have lost much of its internal diversity. When the populations are allowed to expand again, they will be more genetically uniform than their ancestral populations. For example, the bison herds of today are biologically not the same in terms of their genetic diversity as the bison herds of the early 18th century.

Population geneticists have developed mathematical formulae to express a genetically effective population size. These explain the genetic effects on populations which have passed through a 'bottleneck' of a small population size, such as the North American bison or African cheetah. The resultant inbreeding may have a number of detrimental effects such as lowered fertility and increased susceptibility to disease. This is termed 'inbreeding depression'.

The effects of small population size depend on the breeding system of the species and the duration of the bottleneck. If the bottleneck lasts for many generations, or population recovery is very slow, a great deal of variation can be lost. The converse, 'outbreeding depression', occurs when species become genetically differentiated across their range, and then individuals from different parts of the range breed.

Genetic differentiation within species occurs as a result of either sexual reproduction, in which genetic differences from individuals may be combined in their offspring to produce new combinations of genes, or from mutations which cause changes in the DNA. The significance of genetic diversity is often highlighted with reference to global agriculture and food security. This stresses the reliance of the majority of the world's human population on a small number of staple food species, which in turn rely on supply of genes from their wild relatives to supply new characteristics, for example to improve resistance to pests and diseases.

Species Diversity

Species are regarded as populations within which gene flow occurs under natural conditions. Within a species, all normal individuals are capable of breeding with the other individuals of the opposite sex belonging to the same species, or at least they are capable of being genetically linked with them through chains of other breeding individuals.

By definition, members of one species do not breed freely with members of other species. Although this definition works well for many animal and plant species, it is more difficult to delineate species in populations where hybridisation, or self-fertilisation or parthenogenesis occur. Arbitrary divisions must be made, and indeed this is an area where scientists often disagree.

New species may be established through the process of polyploidy, the multiplication of the number of gene-bearing chromosomes, or more commonly, as a result of geographic speciation. This is the process by which isolated populations diverge by evolution as a result of being subjected to different environmental conditions. Over a long period of time, differences between populations may become great enough to reduce interbreeding and eventually populations may be able to co-exist as newly formed, separate species. Within the hierarchical system used by scientists to classify organisms, species represent the lowest rung on this ladder of classification. In descending order, the main categories, or taxa, of living things are:

— Kingdom

— Phylum

— Class

— Order

— Family

— Genus

— Species

We do not know the true number of species on earth, even to the nearest order of magnitude. Wilson estimates that the absolute number of species falls between 5 and 30 million, although some scientists have put forward even higher estimates, up to 50 million. At present approximately 1.4 million living species of all kinds of organisms have been described.

The best catalogued groups include vertebrates and flowering plants, with other groups relatively under-researched, such as lichens, bacteria, fungi and roundworms. Likewise, some habitats are better researched than others, and coral reefs, deep ocean floor and tropical soils are not well studied. This lack of knowledge has considerable implications for the economics of biodiversity conservation, particularly in defining priorities for cost-effective conservation interventions.

The single most obvious pattern in the global distribution of species is that overall species richness increases with decreasing latitude. Not only does this apply as a general rule, it also holds within the great majority of higher taxa, at order level or higher. However, this overall pattern masks a large number of minor trends. Species richness in particular taxonomic groups, or in particular habitats, may show no significant latitudinal variation, or may actually decrease with decreasing latitudes.

In addition, in terrestrial ecosystems, diversity generally decreases with increasing altitude. This phenomenon is most apparent at extremes of altitude, with the highest regions at all latitudes having very low species diversity. In terms of marine systems, depth is the analogue of altitude in terrestrial systems and biodiversity tends to be negatively correlated with depth. Gradients and changes in species richness are also noticeably correlated to precipitation, nutrient levels and salinity, as well as other climatic variations and available energy.

Ecosystem Diversity

Ecosystem diversity relates to the variety of habitats, biotic communities and ecological processes in the biosphere as well as the diversity within ecosystems. Diversity can be described at a number of different levels and scales:

— Functional diversity is the relative abundance of functionally different kinds of organisms.

— Community diversity is the number sizes and spatial distribution of communities, and is sometimes referred to as patchiness.

— Landscape diversity is the diversity of scales of patchiness.

No simple relationship exists between the diversity of an ecosystem and ecological processes such as productivity, hydrology, and soil generation. Neither does diversity correlate neatly with ecosystem stability, its resistance to disturbance and its speed of recovery. There is no simple relationship within any ecosystem between a change in its diversity and the resulting change in the system's processes.

For example, the loss of a species from a particular area or region may have little or no effect on net primary productivity if competitors take its place in the community. The converse may be true in other cases. For example, if herbivores such as zebra and wildebeest are removed from the African savanna, net primary productivity of the ecosystem decreases. Despite these anomalies, Reid and Miller suggest six general rules of ecosystem dynamics which link environmental changes, biodiversity and ecosystem processes.

1. The mix of species making up communities and ecosystems changes continually.

2. Species diversity increases as environmental heterogeneity or the patchiness of a habitat does, but increasing patchiness does not necessarily result in increased species richness.
3. Habitat patchiness influences not only the composition of species in an ecosystem, but also the interactions among species.
4. Periodic disturbances play an important role in creating the patchy environments that foster high species richness. They help to keep an array of habitat patches in various successional states.
5. Both size and isolation of habitat patches can influence species richness, as can the extent of the transition zones between habitats. These transitional zones, or 'ecotones', support species which would not occur in continuous habitats. In temperate zones, ecotones are often more species rich than continuous habitats, although the reverse may be true in tropical forests.
6. Certain species have disproportionate influences on the characteristics of an ecosystem. These include keystone species, whose loss would transform or undermine the ecological processes or fundamentally change the species composition of the community.

The discussion has shown how biodiversity is a very complex and all-embracing concept, which can be interpreted and analysed on a number of levels and scales. The next section examines some approaches to measuring these concepts.

Distribution

Biodiversity is not evenly distributed, rather it varies greatly across the globe as well as within regions. Among other factors, the diversity of all living things (biota) depends on temperature, precipitation, altitude, soils, geography and the presence of other species. The study of the spatial distribution of organisms, species, and ecosystems, is the science of biogeography.

Diversity consistently measures higher in the tropics and in other localized regions such as the Cape Floristic Region and lower in polar regions generally. In 2006 many species were formally classified as rare or endangered or threatened; moreover, scientists have estimated that millions more species are at risk which have not been formally recognized. About 40 percent of the 40,177 species assessed using the IUCN Red List criteria are now listed as threatened with extinction—a total of 16,119.

Terrestrial biodiversity is up to 25 times greater than ocean biodiversity. Although a recent discovered method put the total number of species on Earth at 8.7 million of which 2.1 million were estimated to live in the ocean.

Latitudinal Gradients

Generally, there is an increase in biodiversity from the poles to the tropics. Thus localities at lower latitudes have more species than localities at higher latitudes. This is often referred to as the latitudinal gradient in species diversity. Several ecological mechanisms may contribute to the gradient, but the ultimate factor behind many of them is the greater mean temperature at the equator compared to that of the poles.

Even though terrestrial biodiversity declines from the equator to the poles, some studies claim that this characteristic is unverified in aquatic ecosystems, especially in marine ecosystems. The latitudinal distribution of parasites does not follow this rule.

Hotspots

A biodiversity hotspot is a region with a high level of endemic species. Hotspots were first named in 1988 by Dr. Sabina Virk. Many hotspots have large nearby human populations. While hotspots are spread all over the world, the majority are forest areas and most are located in the tropics.

Brazil's Atlantic Forest is considered one such hotspot, containing roughly 20,000 plant species, 1,350 vertebrates, and millions of insects, about half of which occur nowhere else. The island of Madagascar, particularly the unique Madagascar dry deciduous forests and lowland rainforests, possess a high ratio of endemism. Since the island separated from mainland Africa 65 million years ago, many species and ecosystems have evolved independently. Indonesia's 17,000 islands cover 735,355 square miles (1,904,560 km2) contain 10% of the world's flowering plants, 12% of mammals and 17% of reptiles, amphibians and birds—along with nearly 240 million people. Many regions of high biodiversity and/or endemism arise from specialized habitats which require unusual adaptations, for example alpine environments in high mountains, or Northern European peat bogs.

Accurately measuring differences in biodiversity can be difficult. Selection bias amongst researchers may contribute to biased empirical research for modern estimates of biodiversity. In 1768 Rev. Gilbert White succinctly observed of his Selborne, Hampshire "all nature is so full, that that district produces the most variety which is the most examined."

Evolution of Biodiversity

Biodiversity is the result of 3.5 billion years of evolution. The origin of life has not been definitely established by science, however some evidence suggests that life may already have been well-established only a few hundred million years after the formation of the Earth. Until approximately 600 million years ago, all life consisted of archaea, bacteria, protozoans and similar single-celled organisms.

The history of biodiversity during the Phanerozoic (the last 540 million years), starts with rapid growth during the Cambrian explosion—a period during which nearly every phylum of multicellular organisms first appeared. Over the next 400 million years or so, invertebrate diversity showed little overall trend, and vertebrate diversity shows an overall exponential trend. This dramatic rise in diversity was marked by periodic, massive losses of diversity classified as mass extinction events. A significant loss occurred when rainforests collapsed in the carboniferous. The worst was the Permo-Triassic extinction, 251 million years ago. Vertebrates took 30 million years to recover from this event.

The fossil record suggests that the last few million years featured the greatest biodiversity in history. However, not all scientists support this view, since there is uncertainty as to how strongly the fossil record is biased by the greater availability and preservation of recent geologic sections. Some scientists believe that corrected for sampling artifacts, modern biodiversity may not be much different from biodiversity 300 million years ago., whereas others consider the fossil record reasonably reflective of the diversification of life. Estimates of the present global macroscopic species diversity vary from 2 million to 100 million, with a best estimate of somewhere near 9 million, the vast majority arthropods. Diversity appears to increase continually in the absence of natural selection.

The existence of a "global carrying capacity", limiting the amount of life that can live at once, is debated, as is the question of whether such a limit would also cap the number of species. While records of life in the sea shows a logistic pattern of growth, life on land (insects, plants and tetrapods)shows an exponential rise in diversity. As one author states, "Tetrapods have not yet invaded 64 per cent of potentially habitable modes, and it could be that without human influence the ecological and taxonomic diversity of tetrapods would continue to increase in an exponential fashion until most or all of the available ecospace is filled."

On the other hand, changes through the Phanerozoic correlate much better with the hyperbolic model (widely used in population biology, demography and macrosociology, as well as fossil biodiversity) than with exponential and logistic models. The latter models imply that changes in diversity are guided by a first-order positive feedback (more ancestors, more descendants) and/or a negative feedback arising from resource limitation. Hyperbolic model implies a second-order positive feedback. The hyperbolic pattern of the world population growth arises from a second-order positive feedback between the population size and the rate of technological growth. The hyperbolic character of biodiversity growth can be similarly accounted for by a feedback between diversity and community structure complexity. The similarity between the curves of biodiversity and human population probably comes from the fact that both are derived from the interference of the hyperbolic trend with cyclical and stochastic dynamics.

Most biologists agree however that the period since human emergence is part of a new mass extinction, named the Holocene extinction event, caused primarily by the impact humans are having on the environment. It has been argued that the present rate of extinction is sufficient to eliminate most species on the planet Earth within 100 years.

New species are regularly discovered (on average between 5–10,000 new species each year, most of them insects) and many, though discovered, are not yet classified (estimates are that nearly 90% of all arthropods are not yet classified). Most of the terrestrial diversity is found in tropical forests and in general, land has more species than the ocean; some 8.7 million species may exists on Earth, of which some 2.1 million live in the ocean

The Essential Role of Biodiversity

The impact of humans on the natural environment is significant and growing. There are currently well over six billion people on the planet; there will likely be nine billion by mid-century. Each person has the right to adequate clean water, food, shelter and energy, the provision of which has profound ecological implications.

Human needs multiplied by a growing world population translate into increasing, and unprecedented, demands on the planet's productive capacity. The growing appetite for consumer goods and services beyond the necessities of survival and the wasteful consumption of available resources by the more privileged segment of global society are exacerbating the strain on the Earth, with consequences for all. As demographic pressures and consumption levels increase, biodiversity decreases, and the ability of the natural world to continue delivering the goods and services on which humanity ultimately depends may be undermined.

Biodiversity underpins ecosystem functioning. The services provided by healthy ecosystems, in turn, are the foundation for human well-being. These ecosystem services not only deliver the basic material needs for survival, but also underlie other aspects of a good life, including health, security, good social relations and freedom of choice. The Millennium Ecosystem Assessment examined the state of 24 services that make a direct contribution to human well-being.

The Assessment concludes that 15 of 24 are in decline, including provision of fresh water, marine fishery production, the number and quality of places of spiritual and religious value, the ability of the atmosphere to cleanse itself of pollutants, natural hazard regulation, pollination, and the capacity of agricultural ecosystems to provide pest control.

By disrupting ecosystem functions, biodiversity loss makes ecosystems more vulnerable to shocks and disturbances, less resilient, and less able to supply humans with needed services. The damage to coastal communities from foods and storms, for example, can increase dramatically following conversion of wetland habitats, as the natural protection offered by these ecosystems against wave action, tidal surge, and water runoff from land is compromised. Recent natural disasters underline this reality.

Healthy ecosystems are critical to human wellbeing at all times, not only in times of catastrophe. For example, inland wetlands are the principal source of renewable fresh water for human use, storing water but also purifying it through the removal of excess nutrients and other pollutants. Disruption of wetland purification processes can have devastating impacts at the source and further downstream.

The loss of wetlands in the Mississippi watershed of the United States, for example, combined with high nutrient loads from intensive agriculture in the region, has contributed to the creation of a low-oxygen "dead zone", incapable of supporting animal life, which extends, on average at midsummer, some 16,000 square kilometres into the Gulf of Mexico.

The consequences of biodiversity loss and ecosystem disruption are often harshest for the rural poor, who depend most immediately upon local ecosystem services for their livelihoods and who are often the least able to access or afford substitutes when these become degraded. In daily life, rural households depend, to varying degrees, on farming, fishing, hunting and the harvest of wild products to help meet their subsistence and cash needs, complementing this environmental income with outside sources of earnings, such as wage labour or remittances.

In times of crisis-during a drought or economic recession, for example-even those households not normally reliant on environmental income can turn to wild products as a last resort. Ecosystems then serve the additional function of social safety nets, insuring families against absolute poverty and starvation. The marginal position of rural communities in society often allows more powerful interests to capture ecosystem benefits for private gain, frequently through the conversion of ecosystems to other uses. Although studies are few, in every case examined where the total economic value of ecosystems under alternative management regimes were compared, managing the ecosystem more sustainably yielded greater total benefits than conversion.

In one of these studies, for instance, intact mangrove ecosystems along Thailand's coast were found to provide substantial benefits to society as a source of timber and non-timber forest products, in the production of charcoal, and by enhancing offshore fisheries and providing storm protection. When mangroves were converted to make way for private shrimp farms, these societal benefits fell to almost zero. Conversion of the natural ecosystem proceeded nonetheless, in part because those individuals standing to gain immediate private benefits did not have to bear the costs associated with the loss of ecosystem services.

In some cases, government subsidies can exaggerate the private benefits of conversion, as ecosystems are degraded at public expense. The end result for the poor is further disenfranchisement. Garnering the political will to halt ecosystem degradation will depend on clearly demonstrating to policy makers and society at large the full contribution made by ecosystems to national economies. A recent World Bank report estimates that natural capital, even when defined narrowly, constitutes a quarter (26%) of the total wealth of low-income countries.

The report also suggests that better management of ecosystems and natural resources will be key to sustaining development while nations build other forms of wealth. Specific examples of the economic value derived from biodiversity are also available, and are increasing in number. However, a more profound rethinking of economic growth, and how it is measured, is also needed. Current measures of economic wealth, such as the gross domestic product (GDP), do not reflect the total economic value of ecosystems, and mistakenly treat nature's goods and services as free to use and limitless in abundance.

As a result, countries that fell their forests for timber exports, dynamite reefs for fish, and degrade their land as a result of unsustainable agriculture can appear to be getting richer in the short-term. Applying better valuation methods to national economies, as indicated in the case study on conversion of mangrove to aquaculture in Thailand, would reveal that for many countries, and in a number of sectors, economic gains as traditionally measured are illusory.

World Bank figures suggest that, per capita, most low-income countries have experienced declines in both total and natural capital, jeopardizing both economic growth and the achievement of the Millennium Development Goals. In fact, the Millennium Ecosystem Assessment has already confirmed that the real costs of biodiversity loss pose a significant barrier to meeting the MDGs.

Although policy-makers have generally focused narrowly on the contribution of biodiversity conservation and sustainable use to the achievement of Goal 7 ("Ensure environmental sustainability"), the wider role of ecosystem services in supporting livelihoods and human well-being reveals biodiversity to be the foundation for all development, and hence for meeting each of the Millennium Development Goals. Studies of food security and nutrition, for instance, have shown the importance of agricultural biodiversity to the elimination of hunger and malnutrition.

In terms of human health, biodiversity also has a recognised role in controlling vector-based diseases and providing the natural sources of many traditional medicines and modern pharmaceutical drugs. The challenge ahead of us lies in the fact that a number of the actions that could be implemented most quickly to promote economic growth and reduce hunger and poverty are harmful to biodiversity, at least in the short- to medium-term, and could undermine the sustainability of any development gains. Recognising the trade-offs and synergies that exist between poverty alleviation, biodiversity conservation and sustainable use will therefore be essential to achieving many of the targets of the Millennium Development Goals.

There are important additional reasons to care about the loss of biodiversity, quite apart from nature's immediate usefulness to humankind. Many would argue that every life form has an intrinsic right to exist. Species alive today are thousands to millions of years old and have each travelled unique evolutionary paths, never to be repeated, in order to reach their present form. We must also recognise the right of future generations to inherit, as we have, a planet thriving with life, and that continues to afford opportunities to reap the economic, cultural and spiritual benefits of nature.

Measurement of Biodiversity

A better understanding of biodiversity can be obtained when we examine exactly what we measure in order to assess biological diversity. However, this also serves to highlight further the range of interpretations, and the importance placed on different hierarchical levels of biodiversity by scholars of different disciplines, and by policy makers. Reid et al., have commented that there is even now no clear consensus about how biodiversity should be measured. Indeed, debates on the measurement of biodiversity have filled a substantial part of the ecological literature since the 1950s.

This lack of consensus also has important implications for the economics of biodiversity conservation. At its most basic level, any measure of cost-effectiveness used to guide investments in conservation must have some index or set of indices of biodiversity change. In the following sections, some aspects of measurement of biodiversity are examined, distinguishing the same components of biodiversity: genetic diversity, species diversity and ecosystem diversity.

Measurement of Genetic Diversity

The analysis and conceptualisation of differences within and among populations is in principle identical regardless of whether we are considering a 'population' to be a local collection of individuals, geographical race, subspecies, species, or higher taxonomic group. Genetic differences can be measured in terms of phenotypic traits, allelic frequencies or DNA sequences.

Phenetic diversity

Phenetic diversity is based on measures of phenotypes, individuals which share the same characteristics. This method avoids examination of the underlying allelic structure. It is usually concerned with measurement of the variance of a particular trait, and often involves readily measurable morphological and physiological characteristics. Phenetic traits can be easily measured, and their ecological or practical utility is either obvious or can be readily inferred. However, their genetic basis is often difficult to assess, and standardised comparisons are difficult when populations or taxa are measured for qualitatively different traits.

Allelic diversity

The same gene can exist in a number of variants and these variants are called alleles. Measures of allelic diversity require knowledge of the allelic composition at individual loci. This information is generally obtained using protein electrophoresis, which analyses the migration of enzymes under the influence of electric field. Allelic diversity may be measured at the individual level, or at the population level.

In general, the more alleles, the more equitable their frequencies, and the more loci that are polymorphic, the greater the genetic diversity. Average expected heterozygosity is commonly used as an overall measure. A number of different indices and coefficients can be applied to the measurements to assess genetic distance. The detection of allelic variation by electrophoresis has the advantage that it can be precisely quantified to provide comparative measures of genetic variation. However, the disadvantages are that it may not be representative of variation in the genome as a whole, and does not take account of functional significance or selective importance of particular alleles.

Sequence variation

A portion of DNA is sequenced using the polymerase chain reaction technique (PCR). This technique means that only a very small amount of material, perhaps one cell, is required to obtain the DNA sequence data, so that only a drop of blood or single hair is required as a sample. Closely related species may share 95 per cent or more of their nuclear DNA sequences, implying a great similarity in the overall genetic information.

Measurement of Species Diversity

Species diversity is a function of the distribution and abundance of species. Often, species richness — the number of species within a region or given area — is used almost synonymously with species diversity. However, technically, species diversity includes some consideration of evenness of species abundances. Let us first consider species richness as a proxy measure of species diversity.

In its ideal form, species richness would consist of a complete catalogue of all species occurring in the area under consideration, but this is not usually possible unless it is a very small area. Species richness measures in practice, therefore, tend to be based on samples. Such samples consist of a complete catalogue of all organisms within a taxa found in a particular area, or of a measure of species density in a given sample plot, or a numerical species richness defined as the number of species per specified number of individuals or biomass.

A more informative measure of diversity would also incorporate the 'relatedness' of the species in a fauna. Using a measure of species richness would imply that a region containing many closely related species would be preferred to one containing a fractionally smaller number of distantly related or genealogically unrelated species. Alternative measures being developed, augment species richness with measures of the degree of genealogical difference. Derived from cladistic methods, these measures include the weighting of close-to-root species, higher-taxon richness, spanning-tree length and taxonomic dispersion.

Close-to-root species and higher-taxon richness explicitly use polarity from the root of the cladogram to weight higher-ranking taxa or 'relic' species as distinct survivors of long-independent lineages and original conduits of genetic information. In contrast, spanning tree length and taxonomic dispersion are more general tree measures of sub-tree 'representativeness'. Polarity from the root of the tree is less important than the amount of the cladogram represented by a fauna or the choice of a fauna to evenly cover the diversity of subgroups found in the cladogram.

There is considerable disagreement as to which measure best characterises the pattern of difference in the popular concept of biodiversity, although there is considerable support for taxonomic dispersion as a method of selecting faunas which most evenly represent a variety of cladogram sub-groups. For the time being, difficulties in actual implementation of cladistic measures suggest reliance on cruder indicators of richness of genera or families for rapid assessment of species diversity.

Measurement of Community Diversity

Many environmentalists and ecologists put emphasis on conservation of biodiversity at the community level. There are a number of factors which make measurement and assessment of diversity at this level more nebulous and less clearly defined. Many different 'units' of diversity are involved at the supra-species level, including the pattern of habitats in the community, relative abundance of species, age structure of populations, patterns of communities on the landscape, trophic structure, and patch dynamics.

At these levels, unambiguous boundaries delineating units of biodiversity do not exist. By conserving biodiversity at the ecosystem level, not only are the constituent species preserved, but also the ecosystem functions and services protected. These include pollutant cycling, nutrient cycling, climate control, as well as non-consumptive recreation, scientific and aesthetic values.

Given the complexities of defining biodiversity at community or ecosystem level already described, there is a range of different approaches to measuring ecosystem diversity. As Reid et al explain, any number of community attributes are components of biodiversity and may deserve monitoring for specific objectives. There are several generic measures of community level diversity. These include biogeographical realms or provinces, based on the distribution of species, and ecoregions or ecozones, based on physical attributes such as soils and climate.

These definitions may differ according to scale. For example, the world has been divided into biogeographical provinces, or more fine-grained classifications which may be more useful for policy-making. More policy orientated measures include the definition of 'hotspots', based on the number of endemic species, and 'megadiversity' states. These concepts will be discussed in the context of using indicators for assessing and monitoring biodiversity.

Biodiversity Loss

Estimates of precise rates of loss of biological diversity are hampered by the absence of any baseline measurement. However, from evidence of island habitats it seems likely that the expansion of the human niche by various forms of conversion is geometrically related to extinctions. Further recent evidence from observation of potential 'indicator' species such as amphibians and birds provides some indication of accelerated loss in excess of historical or background rates.

Table 1 shows some estimates of current rates of species extinction based on extrapolations of human land use trends related to species area curves which are the basis of island biogeography. Over the next century the projected loss of species might be expected to be as high as 20 to 50 per cent of the world's totals which represents a rate between 1000 to 10000 times the historical rate of extinction. The rate of loss is outstripping the natural regenerative capacity of evolution to throw up new or evolved species. The extinction 'outputs' far exceed the speciation 'inputs'.

Table 1. Estimates of the current rates of species extinction

Estimate of loss of species (%)	*Basis*
33–50 by year 2000	Forest area loss
50 by year 2000	Forest area loss
25–30 in 21st century	Forest area loss
33 in 21st century	Forest area loss

The potential effects of accelerated extinction and depletion of the genetic base may be discerned over varying time horizons. In the long term, processes of natural selection and evolution may be dependent on a diminished resource base, simply because fewer species are being born. The implications of species depletion for the integrity of many vital ecosystems are far from clear.

The possible existence of depletion thresholds, associated system collapse, and huge discontinuities in related social cost functions, are potentially the worst outcome in any reasonable human time horizon. Such scenarios are indicative of the links between ecosystem integrity and economic well-being. More immediately, the impoverishment of biological resources in many countries might also be regarded as an antecedent to a decline in community or cultural diversity, indices of which are provided in diet, medicine, language and social structure.

At least four questions emerge from the scientific uncertainty surrounding species loss.

1. What is the number of species from which to measure current rates of loss and the detection of this rate, allowing for background evolutionary turn-over?

2. How sound are the principals and predictions of island biogeography and by how much are current extinction estimates (probably) understated?
3. Given the likely time horizons at issue, can we be concerned with the perversion of the evolutionary processes as opposed to the immediacy of system thresholds and flips?
4. What is the potential for using indicator species or a more sophisticated index to guide conservation efforts, and is there any scientific consensus on appropriate species or ecosystems to be used?

The need to pursue cost-effective investment interventions in biodiversity conservation has added considerable urgency to these issues, and the indicators debate in particular. Reliance on pivotal keystone or umbrella species is appealing but crude. Similarly focusing on wider taxonomic groups or ecosystem functions provides few indications of the likelihood of successful interventions, given wider socio-economic pressures on wilderness ecosystems and protected areas.

Criteria such as species sensitivity to habitat disruption or poor reproductive capacity can be combined with other socio-economic data such as population density, deforestation or figures on conservation investment expenditure, to provide some indication of where species are threatened.

However, assuming some consensus definition of threat raises the issue of whether funding is most effectively directed to those areas most under threat, or away from them entirely in favour of areas with a higher likelihood of success. This in turn implies some objective assessment of a 'successful' intervention. Given that no species can be saved indefinitely, the objective decision criterion becomes the extra cost of an increment to the probability of survival.

Understandably the development of investment criteria designed to maximise diversity per dollar and incorporating a composite threat indicator is likely to take time. Inevitable data restrictions are certain to further complicate what is already a contentious exercise. Nevertheless, two facts are clear. At some point a consensus measurement of biodiversity is required to guide the investment of scarce funds. The resulting index may seem arbitrary and will inevitably contravene some section of scientific opinion but will be necessary to provide a general direction for biodiversity investment.

Moreover, using this index, any cost-effective system of area triage will necessarily require some consideration of complementarity of resulting faunal designations. In other words, the selection of successive areas for protection, ideally needs to be based on the incremental complement to diversity afforded by the last fauna until the complement is reduced to zero. This is clearly a massive undertaking, requiring precise taxonomic inventories and as much socio-economic information as dictated by the guiding index.

Biodiversity Conservation

In summer 1992 in Rio de Janeiro, the world's nations agreed a global Convention on Biological Diversity. It aims to protect the world's biological resources from further erosion or, at least, to slow that rate of erosion down. Yet the rate of erosion of biodiversity is increasing. Despite the Convention, the need to demonstrate the importance of conservation remains as strong as it ever was, perhaps stronger. One aspect of the process of changing government and popular perceptions about biological resources is to show that the sustainable use of biodiversity has positive economic value, and that this economic value will often be higher than the value of alternative resource uses which threaten biodiversity.

Recognition of a broader total economic valuation of natural assets can be instrumental in altering decisions about their use, particularly in investment and land-use decisions which present a clear choice between destruction or conservation. Such decisions are being faced in both developed and developing countries, where a host of competing social and economic claims increasingly conflict with the resource demands of area protection.

Greater understanding of the functioning of natural ecosystems combined with enhanced valuation techniques are an increasing influence on national conservation strategies, while international and multilateral initiatives emphasize the global dimension to the issue of biodiversity loss. We argue that addressing the economic causes of biodiversity loss is extremely important if the world really does want to slow down the erosion of its biological resources. Much of the biodiversity that needs saving resides in the developing world.

Since biodiversity conservation is not, understandably, a priority for the developing world, the resources needed for conservation must come from the North, while the political commitment must come from the South and

North alike. However we would like the world to be, the brute fact is that only policies which offer mutual self-interested gains to both North and South stand a chance of succeeding. In the longer term we may hope for changes of attitudes and priorities in the world generally, especially as incomes rise in the South. But relying on such changes to bring about conservation is foolhardy and counterproductive. That is why the economic approach matters. It does emphasise mutual economic gain as the foundation for the solution to the biodiversity problem.

REFERENCES

Leveque, C. & J. Mounolou (2003) *Biodiversity*. New York: John Wiley.

Margulis, L., Dolan, Delisle, K., Lyons, C. *Diversity of Life: The Illustrated Guide to the Five Kingdoms*. Sudbury: Jones & Bartlett Publishers.

Novacek, M. J. (ed.) (2001) *The Biodiversity Crisis: Losing What Counts*. New York: American Museum of Natural HistoryBooks.

Pereira, H. M.; Navarro, L. M.; Martins, I. S. S. (2012). "Global Biodiversity Change: The Bad, the Good, and the Unknown".*Annual Review of Environment and Resources* 37: 25.

2

Habitat Loss

Habitat loss—due to destruction, fragmentation or degradation of habitat—is the primary threat to the survival of biodiversity. When an ecosystem has been dramatically changed by human activities—such as agriculture, oil and gas exploration, commercial development or water diversion—it may no longer be able to provide the food, water, cover, and places to raise young. Every day there are fewer places left that wildlife can call home.

There are three major kinds of habitat loss:

1. *Habitat destruction*: A bulldozer pushing down trees is the iconic image of habitat destruction. Other ways that people are directly destroying habitat, include filling in wetlands, dredging rivers, mowing fields, and cutting down trees.
2. *Habitat fragmentation:* Much of the remaining terrestrial wildlife habitat in the U.S. has been cut up into fragments by roads and development. Aquatic species' habitat has been fragmented by dams and water diversions. These fragments of habitat may not be large or connected enough to support species that need a large territory in which to find mates and food. The loss and fragmentation of habitat make it difficult for migratory species to find places to rest and feed along their migration routes.
3. *Habitat degradation*: Pollution, invasive species and disruption of ecosystem processes (such as changing the intensity of fires in an ecosystem) are some of the ways habitats can become so degraded that they no longer support native wildlife.

Habitat loss is where humans or other species take over or destroy the home of another. In this process, the organisms that previously used the site are displaced or destroyed, reducing biodiversity. Habitat loss by human activity mainly for the purpose of harvesting natural resources for industry production and urbanization. Clearing habitats for agriculture is the principal cause of habitat loss. Other important causes of habitat loss include mining, logging, trawling and urban sprawl. Habitat loss is currently ranked as the primary cause of species extinction worldwide. It is a process of natural environmental change that may be caused by habitat fragmentation, geological processes, climate change or by human activities such as the introduction of invasive species, ecosystem nutrient depletion and other human activities mentioned below.

Impacts on Organisms

In the simplest term, when a habitat is destroyed, the plants, animals, and other organisms that occupied the habitat have a reduced carrying capacity so that populations decline and extinction becomes more likely. Perhaps the greatest threat to organisms and biodiversity is the process of habitat loss. Temple (1986) found that 82% of endangered bird species were significantly threatened by habitat loss. Endemic organisms with limited ranges are most affected by habitat destruction, mainly because these organisms are not found anywhere else within the world and thus, have less chance of recovering. Many endemic organisms have very specific requirements for their survival that can only be found within a certain ecosystem, resulting in their extinction. Extinction may take place very long after the destruction of habitat, however, a phenomenon known as extinction debt. Habitat destruction can also decrease the range of certain organism populations. This can result in the reduction of genetic diversity and perhaps the production of infertile youths, as these organisms would have a higher possibility of mating with related organisms within their population, or different species. One of the most famous examples is the impact upon China's Giant Panda, once found across the nation. Now it's only found in fragmented and isolated regions in the south-west of the country, as a result of widespread deforestation in the 20th Century.

Geography of Habitat Destruction

Biodiversity hotspots are chiefly tropical regions that feature high concentrations of endemic species and, when all hotspots are combined, may contain over half of the world's terrestrial species. These hotspots are suffering from habitat loss and destruction. Most of the natural habitat on islands and in areas of high human population density has already been destroyed (WRI, 2003). Islands suffering extreme habitat destruction include New Zealand, Madagascar, the Philippines, and Japan. South and east Asia — especially China, India, Malaysia, Indonesia, and Japan — and many areas in West Africa have extremely dense human populations that allow little room for natural habitat. Marine areas close to highly populated coastal cities also face degradation of their coral reefs or other marine habitat. These areas include the eastern coasts of Asia and Africa, northern coasts of South America, and the Caribbean Sea and its associated islands.

Regions of unsustainable agriculture or unstable governments, which may go hand-in-hand, typically experience high rates of habitat destruction. Central America, Sub-Saharan Africa, and the Amazonian tropical rainforest areas of South America are the main regions with unsustainable agricultural practices or government mismanagement.

Areas of high agricultural output tend to have the highest extent of habitat destruction. In the U.S., less than 25% of native vegetation remains in many parts of the East and Midwest. Only 15% of land area remains unmodified by human activities in all of Europe.

Ecosystems

Tropical rainforests have received most of the attention concerning the destruction of habitat. From the approximately 16 million square kilometers of tropical rainforest habitat that originally existed worldwide, less than 9 million square kilometers remain today. The current rate of deforestation is 160,000 square kilometers per year, which equates to a loss of approximately 1% of original forest habitat each year.

Other forest ecosystems have suffered as much or more destruction as tropical rainforests. Farming and logging have severely disturbed at least 94% of temperate broadleaf forests; many old growth forest stands have lost more than 98% of their previous area because of human activities. Tropical deciduous dry forests are easier to clear and burn and are more suitable for agriculture and cattle ranching than tropical rainforests; consequently, less than 0.1% of dry forests in Central America's Pacific Coast and less than 8% in Madagascar remain from their original extents.

Plains and desert areas have been degraded to a lesser extent. Only 10-20% of the world's drylands, which include temperate grasslands, savannas, and shrublands, scrub, and deciduous forests, have been somewhat degraded. But included in that 10-20% of land is the approximately 9 million square kilometers of seasonally dry-lands that humans have converted to deserts through the process of desertification. The tallgrass prairies of North America, on the other hand, have less than 3% of natural habitat remaining that has not been converted to farmland.

Wetlands and marine areas have endured high levels of habitat destruction. More than 50% of wetlands in the U.S. have been destroyed in just the last 200 years. Between 60% and 70% of European wetlands have been completely destroyed. About one-fifth (20%) of marine coastal areas have been highly modified by humans. One-fifth of coral reefs have also been destroyed, and another fifth has been severely degraded by overfishing, pollution, and invasive species; 90% of the Philippines' coral reefs alone have been destroyed. Finally, over 35% mangrove ecosystems worldwide have been destroyed.

Causes of Habitat Loss

Natural Causes

Habitat destruction through natural processes such as volcanism, fire, and climate change is well documented in the fossil record. One study shows that habitat fragmentation of tropical rainforests in Euramerica 300 million years ago led to a great loss of amphibian diversity, but simultaneously the drier climate spurred on a burst of diversity among reptiles.

Human Causes

Habitat destruction caused by humans includes conversion of land to agriculture, urban sprawl, infrastructure development, and other anthropogenic changes to the characteristics of land. Habitat degradation, fragmentation, and pollution are aspects of habitat destruction caused by humans that do not necessarily involve overt destruction of habitat, yet result in habitat collapse. Desertification, deforestation, and coral reef degradation are specific types of habitat destruction for those areas (deserts, forests, coral reefs).

Geist and Lambin assessed 152 case studies of net losses of tropical forest cover to determine any patterns in the proximate and underlying causes of tropical deforestation. Their results, yielded as percentages of the case studies in which each parameter was a significant factor, provide a quantitative prioritization of which proximate and underlying causes were the most significant. The proximate causes were clustered into broad categories of agricultural expansion (96%), infrastructure expansion (72%), and wood extraction (67%). Therefore, according to this study, forest conversion to agriculture is the main land use change responsible for tropical deforestation. The specific categories reveal further insight into the specific causes of tropical deforestation: transport extension (64%), commercial wood extraction (52%), permanent cultivation (48%), cattle ranching (46%), shifting (slash and burn) cultivation (41%), subsistence agriculture (40%), and fuel wood extraction for domestic use (28%). One result is that shifting cultivation is not the primary cause of deforestation in all world regions, while transport extension (including the construction of new roads) is the largest single proximate factor responsible for deforestation.

Drivers of Habitat Loss

While the above-mentioned activities are the proximal or direct causes of habitat destruction in that they actually destroy habitat, this still does not identify why humans destroy habitat. The forces that cause humans to destroy habitat are known as drivers of habitat destruction. Demographic, economic, sociopolitical, scientific and technological, and cultural drivers all contribute to habitat destruction.

Demographic drivers include the expanding human population; rate of population increase over time; spatial distribution of people in a given area (urban versus rural), ecosystem type, and country; and the combined effects of poverty, age, family planning, gender, and education status of people in certain areas. Most of the exponential human population growth worldwide is occurring in or close to biodiversity hotspots. This may explain why human population density accounts for 87.9% of the variation in numbers of threatened species across 114 countries, providing indisputable evidence that people play the largest role in decreasing biodiversity. The boom in human population and migration of people into such species-rich regions are making conservation efforts not only more urgent but also more likely to conflict with local human interests. The high local population density in such areas is directly correlated to the poverty status of the local people, most of whom lacking an education and family planning.

From the Geist and Lambin study described in the previous section, the underlying driving forces were prioritized as follows (with the percent of the 152 cases the factor played a significant role in): economic factors (81%), institutional or policy factors (78%), technological factors (70%), cultural or socio-political factors (66%), and demographic factors (61%). The main economic factors included commercialization and growth of timber markets (68%), which are driven by national and international demands; urban industrial growth (38%); low domestic costs for land, labor, fuel, and timber (32%); and increases in product prices mainly for cash crops (25%). Institutional and policy factors included formal pro-deforestation policies on land development (40%), economic growth including colonization and infrastructure improvement (34%), and subsidies for land-based activities (26%); property rights and land-tenure insecurity (44%); and policy failures such as corruption, lawlessness, or mismanagement (42%). The main technological factor was the poor application of technology in the wood industry (45%), which leads to wasteful logging practices. Within the broad category of cultural and sociopolitical factors are public attitudes and values (63%), individual/household behavior (53%), public unconcern toward forest environments (43%), missing basic values (36%), and unconcern by individuals (32%). Demographic factors were the in-migration of colonizing settlers into sparsely populated forest areas (38%) and growing population density — a result of the first factor — in those areas (25%).

There are also feedbacks and interactions among the proximate and underlying causes of deforestation that can amplify the process. Road construction has the largest feedback effect, because it interacts with—and leads to—the establishment of new settlements and more people, which causes a growth in wood (logging) and food markets. Growth in these markets, in turn, progresses the commercialization of agriculture and logging industries. When these industries become commercialized, they must become more efficient by utilizing larger or more modern machinery that often are worse on the habitat than traditional farming and logging methods. Either way, more land is cleared more rapidly for commercial markets. This common feedback example manifests just how closely related the proximate and underlying causes are to each other.

Impact on Human Population

Habitat destruction vastly increases an area's vulnerability to natural disasters like flood and drought, crop failure, spread of disease, and water contamination. On the other hand, a healthy ecosystem with good management practices will reduce the chance of these events happening, or will at least mitigate adverse impacts.

Agricultural land can actually suffer from the destruction of the surrounding landscape. Over the past 50 years, the destruction of habitat surrounding agricultural land has degraded approximately 40% of agricultural land worldwide via erosion, salinization, compaction, nutrient depletion, pollution, and urbanization. Humans also lose direct uses of natural habitat when habitat is destroyed. Aesthetic uses such as birdwatching, recreational uses like hunting and fishing, and ecotourism usually rely upon virtually undisturbed habitat. Many people value the complexity of the natural world and are disturbed by the loss of natural habitats and animal or plant species worldwide.

Probably the most profound impact that habitat destruction has on people is the loss of many valuable ecosystem services. Habitat destruction has altered nitrogen, phosphorus, sulfur, and carbon cycles, which has increased the frequency and severity of acid rain, algal blooms, and fish kills in rivers and oceans and contributed tremendously to global climate change. One ecosystem service whose significance is becoming more realized is climate regulation.

On a local scale, trees provide windbreaks and shade; on a regional scale, plant transpiration recycles rainwater and maintains constant annual rainfall; on a global scale, plants (especially trees from tropical rainforests) from around the world counter the accumulation of greenhouse gases in the atmosphere by sequestering carbon dioxide through photosynthesis. Other ecosystem services that are diminished or lost altogether as a result of habitat destruction include watershed management, nitrogen fixation, oxygen production, pollination, waste treatment (i.e., the breaking down and immobilization of toxic pollutants), and nutrient recycling of sewage or agricultural runoff.

The loss of trees from the tropical rainforests alone represents a substantial diminishing of the earth's ability to produce oxygen and use up carbon dioxide. These services are becoming even more important as increasing carbon dioxide levels is one of the main contributors to global climate change.

The loss of biodiversity may not directly affect humans, but the indirect effects of losing many species as well as the diversity of ecosystems in general are enormous. When biodiversity is lost, the environment loses many species that provide valuable and unique roles to the ecosystem. The environment and all its inhabitants rely on biodiversity to recover from extreme environmental conditions. When too much biodiversity is lost, a catastrophic event such as an earthquake, flood, or volcanic eruption could cause an ecosystem to crash, and humans would obviously suffer from that. Loss of biodiversity also means that humans are losing animals that could have served as biological control agents and plants that could potentially provide higher-yielding crop varieties, pharmaceutical drugs to cure existing or future diseases or cancer, and new resistant crop varieties for agricultural species susceptible to pesticide-resistant insects or virulent strains of fungi, viruses, and bacteria.

The negative effects of habitat destruction usually impact rural populations more directly than urban populations. Across the globe, poor people suffer the most when natural habitat is destroyed, because less natural habitat means less natural resources per capita, yet wealthier people and countries simply have to pay more to continue to receive more than their per capita share of natural resources.

Another way to view the negative effects of habitat destruction is to look at the opportunity cost of keeping an area undisturbed. In other words, what are people losing out on by taking away a given habitat? A country may increase its food supply by converting forest land to row-crop agriculture, but the value of the same land may be much larger when it can supply natural resources or services such as clean water, timber, ecotourism, or flood regulation and drought control.

Outlook

The rapid expansion of the global human population is increasing the world's food requirement substantially. Simple logic instructs that more people will require more food. In fact, as the world's population increases dramatically, agricultural output will need to increase by at least 50%, over the next 30 years. In the past, continually moving to new land and soils provided a boost in food production to appease the global food demand. That easy fix will no longer be available, however, as more than 98% of all land suitable for agriculture is already in use or degraded beyond repair.

The impending global food crisis will be a major source of habitat destruction. Commercial farmers are going to become desperate to produce more food from the same amount of land, so they will use more fertilizers and less concern for the environment to meet the market demand. Others will seek out new land or will convert other land-uses to agriculture. Agricultural intensification will become widespread at the cost of the environment and its inhabitants. Species will be pushed out of their habitat either directly by habitat destruction or indirectly by fragmentation, degradation, or pollution. Any efforts to protect the world's remaining natural habitat and biodiversity will compete directly with humans' growing demand for natural resources, especially new agricultural lands.

Solutions

In most cases of tropical deforestation, three to four underlying causes are driving two to three proximate causes. This means that a universal policy for controlling tropical deforestation would not be able to address the unique combination of proximate and underlying causes of deforestation in each country. Before any local, national, or international deforestation policies are written and enforced, governmental leaders must acquire a detailed understanding of the complex combination of proximate causes and underlying driving forces of deforestation in a given area or country. This concept, along with many other results about tropical deforestation from the Geist and Lambin study, can easily be applied to habitat destruction in general. Governmental leaders need to take action by addressing the underlying driving forces, rather than merely regulating the proximate causes. In a broader sense, governmental bodies at a local, national, and international scale need to emphasize the following:

- Considering the many irreplaceable ecosystem services provided by natural habitats
- Protecting remaining intact sections of natural habitat
- Educating the public about the importance of natural habitat and biodiversity
- Developing family planning programs in areas of rapid population growth
- Finding ways to increase agricultural output than simply increasing the total land in production

— Preserving habitat corridors to minimize prior damage from fragmented habitats.

— Reduce human population and expansion

References

Barbault, R. and S. D. Sastrapradja. (1995). *Generation, maintenance and loss of biodiversity. Global Biodiversity Assessment,* Cambridge Univ. Press, Cambridge pp. 193–274.

MEA. (2005). *Ecosystems and Human Well-Being. Millennium Ecosystem Assessment.* Island Press, Covelo, CA.

Stein, B. A., L. S. Kutner, and J. S. Adams (eds.). (2000). *Precious Heritage: The Status of Biodiversity in the United States.* Oxford University Press, New York.

White, R. P., S. Murray, and M. Rohweder. (2000). *Pilot Assessment of Global Ecosystems: Grassland Ecosystems.* World Resources Institute, Washington, DC.

3

Deforestation

Deforestation is the removal of a forest or stand of trees where the land is thereafter converted to a non-forest use. Examples of deforestation include conversion of forestland to farms, ranches, or urban use. About half of the world's original forests had been destroyed by 2011, the majority during the previous 50 years. Since 1990 half of the world's rain forests have been destroyed. More than half of the animal and plant species in the world live in tropical forests.

The term deforestation is often misused to describe any activity where all trees in an area are removed. However in temperate climates, the removal of all trees in an area—in conformance with sustainable forestry practices—is correctly described as regeneration harvest. In temperate mesic climates, natural regeneration of forest stands often will not occur in the absence of disturbance, whether natural or anthropogenic. Furthermore, biodiversity after regeneration harvest often mimics that found after natural disturbance, including biodiversity loss after naturally occurring rainforest destruction.

Deforestation occurs for many reasons: trees are cut down to be used or sold as fuel (sometimes in the form of charcoal) or timber, while cleared land is used as pasture for livestock, plantations of commodities and settlements. The removal of trees without sufficient reforestation has resulted in damage to habitat, biodiversity loss and aridity. It has adverse impacts on biosequestration of atmospheric carbon dioxide. Deforestation has also been used in war to deprive an enemy of cover for its forces and also vital resources. A modern example of this was the use of Agent Orange by the

United States military in Vietnam during the Vietnam War. Deforested regions typically incur significant adverse soil erosion and frequently degrade into wasteland.

Disregard or ignorance of intrinsic value, lack of ascribed value, lax forest management and deficient environmental laws are some of the factors that allow deforestation to occur on a large scale. In many countries, deforestation, both naturally occurring and human induced, is an ongoing issue. Deforestation causes extinction, changes to climatic conditions, desertification, and displacement of populations as observed by current conditions and in the past through the fossil record.

Causes of Deforestation

According to the United Nations Framework Convention on Climate Change (UNFCCC) secretariat, the overwhelming direct cause of deforestation is agriculture. Subsistence farming is responsible for 48% of deforestation; commercial agriculture is responsible for 32% of deforestation; logging is responsible for 14% of deforestation and fuel wood removals make up 5% of deforestation.

Experts do not agree on whether industrial logging is an important contributor to global deforestation. Some argue that poor people are more likely to clear forest because they have no alternatives, others that the poor lack the ability to pay for the materials and labour needed to clear forest. One study found that population increases due to high fertility rates were a primary driver of tropical deforestation in only 8% of cases.

Other causes of contemporary deforestation may include corruption of government institutions, the inequitable distribution of wealth and power, population growth and overpopulation, and urbanization. Globalization is often viewed as another root cause of deforestation, though there are cases in which the impacts of globalization have promoted localized forest recovery.

In 2000 the United Nations Food and Agriculture Organization (FAO) found that "the role of population dynamics in a local setting may vary from decisive to negligible," and that deforestation can result from "a combination of population pressure and stagnating economic, social and technological conditions."

The degradation of forest ecosystems has also been traced to economic incentives that make forest conversion appear more profitable than forest conservation. Many important forest functions have no markets, and hence, no economic value that is readily apparent to the forests' owners or the communities that rely on forests for their well-being. From the perspective of the developing world, the benefits of forest as carbon sinks or biodiversity reserves go primarily to richer developed nations and there is insufficient compensation for these services. Developing countries feel that some countries in the developed world, such as the United States of America, cut down their forests centuries ago and benefited greatly from this deforestation, and that it is hypocritical to deny developing countries the same opportunities: that the poor shouldn't have to bear the cost of preservation when the rich created the problem.

Some commentators have noted a shift in the drivers of deforestation over the past 30 years. Whereas deforestation was primarily driven by subsistence activities and government-sponsored development projects like transmigration in countries like Indonesia and colonization in Latin America, India, Java, and so on, during late 19th century and the earlier half of the 20th century. By the 1990s the majority of deforestation was caused by industrial factors, including extractive industries, large-scale cattle ranching, and extensive agriculture.

Environmental Problems

Atmospheric

Deforestation is a contributor to global warming, and is often cited as one of the major causes of the enhanced greenhouse effect. Tropical deforestation is responsible for approximately 20% of world greenhouse gas emissions. According to the Intergovernmental Panel on Climate Change deforestation, mainly in tropical areas, could account for up to one-third of total anthropogenic carbon dioxide emissions. But recent calculations suggest that carbon dioxide emissions from deforestation and forest degradation (excluding peatland emissions) contribute about 12% of total anthropogenic carbon dioxide emissions with a range from 6 to 17%. Deforestation causes carbon dioxide to linger in the atmosphere. As carbon dioxide accrues, it produces a layer in the atmosphere that traps radiation from the sun. The radiation converts to heat which causes global warming, which is better known as the greenhouse effect. Other plants remove carbon (in the form of carbon dioxide) from the atmosphere during the process of photosynthesis and release oxygen back into the atmosphere during normal respiration. Only when actively growing can a tree or forest remove carbon over an annual or longer timeframe. Both the decay and burning of wood releases much of this stored carbon back to the atmosphere. In order for forests to take up carbon, the wood must be harvested and turned into long-lived products and trees must be re-planted. Deforestation may cause carbon stores held in soil to be released. Forests are stores of carbon and can be either sinks or sources depending upon environmental circumstances. Mature forests alternate between being net sinks and net sources of carbon dioxide. In deforested areas, the land heats up faster and reaches a higher temperature, leading to localized upward motions that enhance the formation of clouds and ultimately produce more rainfall. However, according to the Geophysical Fluid Dynamics Laboratory, the models used to investigate remote responses to tropical deforestation showed a broad but mild temperature increase all through the tropical atmosphere. The model predicted <0.2°C warming for upper air at 700 mb and 500 mb. However, the model shows no significant changes in other areas besides the Tropics. Though the model showed no significant changes to the climate in areas other than the Tropics, this may not be the case since the model has possible errors and the results are never absolutely definite.

Reducing emissions from the tropical deforestation and forest degradation (REDD) in developing countries has emerged as new potential to complement ongoing climate policies. The idea consists in providing financial compensations for the reduction of greenhouse gas (GHG) emissions from deforestation and forest degradation".

Rainforests are widely believed by laymen to contribute a significant amount of world's oxygen, although it is now accepted by scientists that rainforests contribute little net oxygen to the atmosphere and deforestation has only a minor effect on atmospheric oxygen levels. However, the incineration and burning of forest plants to clear land releases large amounts of CO_2, which contributes to global warming. Scientists also state that, Tropical deforestation releases 1.5 billion tons of carbon each year into the atmosphere.

Hydrological

The water cycle is also affected by deforestation. Trees extract groundwater through their roots and release it into the atmosphere. When part of a forest is removed, the trees no longer transpire this water, resulting in a much drier climate. Deforestation reduces the content of water in the soil and groundwater as well as atmospheric moisture. The dry soil leads to lower water intake for the trees to extract. Deforestation reduces soil cohesion, so that erosion, flooding and landslides ensue.

Shrinking forest cover lessens the landscape's capacity to intercept, retain and transpire precipitation. Instead of trapping precipitation, which then percolates to groundwater systems, deforested areas become sources of surface water runoff, which moves much faster than subsurface flows. That quicker transport of surface water can translate into flash flooding and more localized floods than would occur with the forest cover. Deforestation also contributes to decreased evapotranspiration, which lessens atmospheric moisture which in some cases affects precipitation levels downwind from the deforested area, as water is not recycled to downwind forests, but is lost in runoff and returns directly to the oceans. According to one study, in deforested north and northwest China, the average annual precipitation decreased by one third between the 1950s and the 1980s.

Trees, and plants in general, affect the water cycle significantly:

— their canopies intercept a proportion of precipitation, which is then evaporated back to the atmosphere (canopy interception);
— their litter, stems and trunks slow down surface runoff;
— their roots create macropores – large conduits – in the soil that increase infiltration of water;
— they contribute to terrestrial evaporation and reduce soil moisture via transpiration;
— their litter and other organic residue change soil properties that affect the capacity of soil to store water.
— their leaves control the humidity of the atmosphere by transpiring. 99% of the water absorbed by the roots moves up to the leaves and is transpired.

As a result, the presence or absence of trees can change the quantity of water on the surface, in the soil or groundwater, or in the atmosphere. This in turn changes erosion rates and the availability of water for either ecosystem functions or human services.

The forest may have little impact on flooding in the case of large rainfall events, which overwhelm the storage capacity of forest soil if the soils are at or close to saturation.

Soil

Undisturbed forests have a very low rate of soil loss, approximately 2 metric tons per square kilometer (6 short tons per square mile). Deforestation generally increases rates of soil erosion, by increasing the amount of runoff and reducing the protection of the soil from tree litter. This can be an advantage in excessively leached tropical rain forest soils. Forestry operations themselves also increase erosion through the development of roads and the use of mechanized equipment.

China's Loess Plateau was cleared of forest millennia ago. Since then it has been eroding, creating dramatic incised valleys, and providing the sediment that gives the Yellow River its yellow color and that causes the flooding of the river in the lower reaches.

Removal of trees does not always increase erosion rates. In certain regions of southwest US, shrubs and trees have been encroaching on grassland. The trees themselves enhance the loss of grass between tree canopies. The bare intercanopy areas become highly erodible. The US Forest Service, in Bandelier National Monument for example, is studying how to restore the former ecosystem, and reduce erosion, by removing the trees.

Tree roots bind soil together, and if the soil is sufficiently shallow they act to keep the soil in place by also binding with underlying bedrock. Tree removal on steep slopes with shallow soil thus increases the risk of landslides, which can threaten people living nearby.

Deforestation and Biodiversity

Deforestation on a human scale results in decline in biodiversity. and on a natural global scale is known to cause the extinction of many species. The removal or destruction of areas of forest cover has resulted in a degraded environment with reduced biodiversity. Forests support biodiversity, providing habitat for wildlife; moreover, forests foster medicinal conservation. With forest biotopes being irreplaceable source of new drugs (such as taxol), deforestation can destroy genetic variations (such as crop resistance) irretrievably.

Since the tropical rainforests are the most diverse ecosystems on Earth and about 80% of the world's known biodiversity could be found in tropical rainforests, removal or destruction of significant areas of forest cover has resulted in a degraded environment with reduced biodiversity.

It has been estimated that we are losing 137 plant, animal and insect species every single day due to rainforest deforestation, which equates to 50,000 species a year. Others state that tropical rainforest deforestation is contributing to the ongoing Holocene mass extinction. The known extinction rates from deforestation rates are very low, approximately 1 species per year from mammals and birds which extrapolates to approximately 23,000 species per year for all species. Predictions have been made that more than 40% of the animal and plant species in Southeast Asia could be wiped out in the 21st century. Such predictions were called into question by 1995 data that show that within regions of Southeast Asia much of the original forest has been converted to monospecific plantations, but that potentially endangered species are few and tree flora remains widespread and stable.

Scientific understanding of the process of extinction is insufficient to accurately make predictions about the impact of deforestation on biodiversity. Most predictions of forestry related biodiversity loss are based on species-area models, with an underlying assumption that as the forest declines species diversity will decline similarly. However, many such models have been proven to be wrong and loss of habitat does not necessarily lead to large scale loss of species. Species-area models are known to overpredict the number of species known to be threatened in areas where actual deforestation is ongoing, and greatly overpredict the number of threatened species that are widespread.

A recent study of the Brazilian Amazon predicts that despite a lack of extinctions thus far, up to 90 percent of predicted extinctions will finally occur in the next 40 years.

Economic Impacts

Damage to forests and other aspects of nature could halve living standards for the world's poor and reduce global GDP by about 7% by 2050, a report concluded at the Convention on Biological Diversity (CBD) meeting in Bonn. Historically, utilization of forest products, including timber and fuel wood, has played a key role in human societies, comparable to the roles of water and cultivable land. Today, developed countries continue to utilize timber for building houses, and wood pulp for paper. In developing countries almost three billion people rely on wood for heating and cooking.

The forest products industry is a large part of the economy in both developed and developing countries. Short-term economic gains made by conversion of forest to agriculture, or over-exploitation of wood products, typically leads to loss of long-term income and long-term biological productivity. West Africa, Madagascar, Southeast Asia and many other regions have experienced lower revenue because of declining timber harvests. Illegal logging causes billions of dollars of losses to national economies annually.

The new procedures to get amounts of wood are causing more harm to the economy and overpower the amount of money spent by people employed in logging. According to a study, "in most areas studied, the various ventures that prompted deforestation rarely generated more than US$5 for every ton of carbon they released and frequently returned far less than US$1". The price on the European market for an offset tied to a one-ton reduction in carbon is 23 euro (about US$35).

Rapidly growing economies also have an effect on deforestation. Most pressure will come from the world's developing countries, which have the fastest-growing populations and most rapid economic (industrial) growth. In 1995, economic growth in developing countries reached nearly 6%, compared with the 2% growth rate for developed countries." As our human population grows, new homes, communities, and expansions of cities will occur. Connecting all of the new expansions will be roads, a very important part in our daily life. Rural roads promote economic development but also facilitate deforestation. About 90% of the deforestation has occurred within 100 km of roads in most parts of the Amazon.

Forest Transition Theory

The forest area change may follow a pattern suggested by the forest transition (FT) theory, whereby at early stages in its development a country is characterized by high forest cover and low deforestation rates (HFLD countries).

Then deforestation rates accelerate (HFHD, high forest cover – high deforestation rate), and forest cover is reduced (LFHD. low forest cover – high deforestation rate), before the deforestation rate slows (LFLD, low forest cover – low deforestation rate), after which forest cover stabilizes and eventually starts recovering. FT is not a "law of nature," and the pattern is influenced by national context (for example, human population density, stage of development, structure of the economy), global economic forces, and government policies. A country may reach very low levels of forest cover before it stabilizes, or it might through good policies be able to "bridge" the forest transition.

FT depicts a broad trend, and an extrapolation of historical rates therefore tends to underestimate future BAU deforestation for counties at the early stages in the transition (HFLD), while it tends to overestimate BAU deforestation for countries at the later stages (LFHD and LFLD).

Countries with high forest cover can be expected to be at early stages of the FT. GDP per capita captures the stage in a country's economic development, which is linked to the pattern of natural resource use, including forests. The choice of forest cover and GDP per capita also fits well with the two key scenarios in the FT:

(i) a forest scarcity path, where forest scarcity triggers forces (for example, higher prices of forest products) that lead to forest cover stabilization; and

(ii) an economic development path, where new and better off-farm employment opportunities associated with economic growth (= increasing GDP per capita) reduce profitability of frontier agriculture and slows deforestation.

Historical Causes

The Carboniferous Rainforest Collapse, was an event that occurred 300 million years ago. Climate change devastated tropical rainforests causing the extinction of many plant and animal species. The change was abrupt, specifically, at this time climate became cooler and drier, conditions that are not favourable to the growth of rainforests and much of the biodiversity within them. Rainforests were fragmented forming shrinking 'islands' further and further apart. This sudden collapse affected several large groups, effects on amphibians were particularly devastating, while reptiles fared better, being ecologically adapted to the drier conditions that followed.

Rainforests once covered 14% of the earth's land surface; now they cover a mere 6% and experts estimate that the last remaining rainforests could be consumed in less than 40 years. Small scale deforestation was practiced by some societies for tens of thousands of years before the beginnings of civilization. The first evidence of deforestation appears in the Mesolithic period. It was probably used to convert closed forests into more open ecosystems favourable to game animals. With the advent of agriculture, larger areas began to be deforested, and fire became the prime tool to clear land for crops. In Europe there is little solid evidence before 7000 BC. Mesolithic foragers used fire to create openings for red deer and wild boar. In Great Britain, shade-tolerant species such as oak and ash are replaced in the pollen record by hazels, brambles, grasses and nettles. Removal of the forests led to decreased transpiration, resulting in the formation of upland peat bogs. Widespread decrease in elm pollen across Europe between 8400–8300 BC and 7200–7000 BC, starting in southern Europe and gradually moving north to Great Britain, may represent land clearing by fire at the onset of Neolithic agriculture.

The Neolithic period saw extensive deforestation for farming land. Stone axes were being made from about 3000 BC not just from flint, but from a wide variety of hard rocks from across Britain and North America as well. They include the noted Langdale axe industry in the English Lake District, quarries developed at Penmaenmawr in North Wales and numerous other locations. Rough-outs were made locally near the quarries, and some were polished locally to give a fine finish. This step not only increased the mechanical strength of the axe, but also made penetration of wood easier. Flint was still used from sources such as Grimes Graves but from many other mines across Europe.

Throughout most of history, humans were hunter gatherers who hunted within forests. In most areas, such as the Amazon, the tropics, Central America, and the Caribbean, only after shortages of wood and other forest products occur are policies implemented to ensure forest resources are used in a sustainable manner.

In ancient Greece, Tjeered van Andel and co-writers summarized three regional studies of historic erosion and alluviation and found that, wherever adequate evidence exists, a major phase of erosion follows, by about 500-1,000 years the introduction of farming in the various regions of Greece, ranging from the later Neolithic to the Early Bronze Age. The thousand years following the mid-first millennium BC saw serious, intermittent pulses of soil erosion in numerous places. The historic silting of ports along the southern coasts of Asia Minor (e.g. Clarus, and the examples of Ephesus, Priene and Miletus, where harbors had to be abandoned because of the silt deposited by the Meander) and in coastal Syria during the last centuries BC.

Easter Island has suffered from heavy soil erosion in recent centuries, aggravated by agriculture and deforestation. Jared Diamond gives an extensive look into the collapse of the ancient Easter Islanders in his book Collapse. The disappearance of the island's trees seems to coincide with a decline of its civilization around the 17th and 18th century. He attributed the collapse to deforestation and over-exploitation of all resources.

The famous silting up of the harbor for Bruges, which moved port commerce to Antwerp, also followed a period of increased settlement growth (and apparently of deforestation) in the upper river basins. In early medieval Riez in upper Provence, alluvial silt from two small rivers raised the riverbeds and widened the floodplain, which slowly buried the Roman settlement in alluvium and gradually moved new construction to higher ground; concurrently the headwater valleys above Riez were being opened to pasturage.

A typical progress trap was that cities were often built in a forested area, which would provide wood for some industry (for example, construction, shipbuilding, pottery). When deforestation occurs without proper replanting, however; local wood supplies become difficult to obtain near enough to remain competitive, leading to the city's abandonment, as happened repeatedly in Ancient Asia Minor. Because of fuel needs, mining and metallurgy often led to deforestation and city abandonment.

With most of the population remaining active in (or indirectly dependent on) the agricultural sector, the main pressure in most areas remained land clearing for crop and cattle farming. Enough wild green was usually left standing (and partially used, for example, to collect firewood, timber and fruits, or to graze pigs) for wildlife to remain viable. The elite's (nobility and higher clergy) protection of their own hunting privileges and game often protected significant woodlands.

Major parts in the spread (and thus more durable growth) of the population were played by monastical 'pioneering' (especially by the Benedictine and Commercial orders) and some feudal lords' recruiting farmers to settle (and become tax payers) by offering relatively good legal and fiscal conditions. Even when speculators sought to encourage towns, settlers needed an agricultural belt around or sometimes within defensive walls. When populations were quickly decreased by causes such as the Black Death or devastating warfare (for example, Genghis Khan's Mongol hordes in eastern and central Europe, Thirty Years' War in Germany), this could lead to settlements being abandoned. The land was reclaimed by nature, but the secondary forests usually lacked the original biodiversity.

From 1100 to 1500 AD, significant deforestation took place in Western Europe as a result of the expanding human population. The large-scale building of wooden sailing ships by European (coastal) naval owners since the 15th century for exploration, colonisation, slave trade–and other trade on the high seas consumed many forest resources. Piracy also contributed to the over harvesting of forests, as in Spain. This led to a weakening of the domestic economy after Columbus' discovery of America, as the economy became dependent on colonial activities (plundering, mining, cattle, plantations, trade, etc.)

In Changes in the Land, William Cronon analyzed and documented 17th-century English colonists' reports of increased seasonal flooding in New England during the period when new settlers initially cleared the forests for agriculture. They believed flooding was linked to widespread forest clearing upstream.

The massive use of charcoal on an industrial scale in Early Modern Europe was a new type of consumption of western forests; even in Stuart England, the relatively primitive production of charcoal has already reached an impressive level. Stuart England was so widely deforested that it depended on the Baltic trade for ship timbers, and looked to the untapped forests of New England to supply the need. Each of Nelson's Royal Navy war ships at Trafalgar (1805) required 6,000 mature oaks for its construction. In France, Colbert planted oak forests to supply the French navy in the future. When the oak plantations matured in the mid-19th century, the masts were no longer required because shipping had changed.

Europeans had lived in the midst of vast forests throughout the earlier medieval centuries. After 1250 they became so skilled at deforestation that by 1500 they were running short of wood for heating and cooking. They were faced with a nutritional decline because of the elimination of the generous supply of wild game that had inhabited the now-disappearing forests, which throughout medieval times had provided the staple of their carnivorous high-protein diet. By 1500 Europe was on the edge of a fuel and nutritional disaster which it was saved in the sixteenth century only by the burning of soft coal and the cultivation of potatoes and maize.

In the 19th century, introduction of steamboats in the United States was the cause of deforestation of banks of major rivers, such as the Mississippi River, with increased and more severe flooding one of the environmental results. The steamboat crews cut wood every day from the riverbanks to fuel the steam engines. Between St. Louis and the confluence with the Ohio River to the south, the Mississippi became more wide and shallow, and changed its channel laterally. Attempts to improve navigation by the use of snagpullers often resulted in crews' clearing large trees 100 to 200 feet (61 m) back from the banks. Several French colonial towns of the Illinois Country, such as Kaskaskia, Cahokia and St. Philippe, Illinois were flooded and abandoned in the late 19th century, with a loss to the cultural record of their archeology.

The wholescale clearance of woodland to create agricultural land can be seen in many parts of the world, such as the Central forest-grasslands transition and other areas of the Great Plains of the United States. Specific parallels are seen in the 20th-century deforestation occurring in many developing nations.

Rates of Global Deforestation

Global deforestation sharply accelerated around 1852. It has been estimated that about half of the Earth's mature tropical forests—between 7.5 million and 8 million km2 (2.9 million to 3 million sq mi) of the original 15 million to 16 million km2 (5.8 million to 6.2 million sq mi) that until 1947 covered the planet—have now been destroyed. Some scientists have predicted that unless significant measures (such as seeking out and protecting old growth forests that have not been disturbed) are taken on a worldwide basis, by 2030 there will only be 10% remaining, with another 10% in a degraded condition. 80% will have been lost, and with them hundreds of thousands of irreplaceable species. Some cartographers have attempted to illustrate the sheer scale of deforestation by country using a cartogram.

Estimates vary widely as to the extent of tropical deforestation. Scientists estimate that one fifth of the world's tropical rainforest was destroyed between 1960 and 1990. They claim that that rainforests 50 years ago covered 14% of the world's land surface, now only cover 5–7%, and that all tropical forests will be gone by the middle of the 21st century.

A 2002 analysis of satellite imagery suggested that the rate of deforestation in the humid tropics (approximately 5.8 million hectares per year) was roughly 23% lower than the most commonly quoted rates. Conversely, a newer analysis of satellite images reveals that deforestation of the Amazon rainforest is twice as fast as scientists previously estimated.

Some have argued that deforestation trends may follow a Kuznets curve, which if true would nonetheless fail to eliminate the risk of irreversible loss of non-economic forest values (for example, the extinction of species).

A 2005 report by the United Nations Food and Agriculture Organization (FAO) estimates that although the Earth's total forest area continues to decrease at about 13 million hectares per year, the global rate of deforestation has recently been slowing. Still others claim that rainforests are being destroyed at an ever-quickening pace. The London-based Rainforest Foundation notes that "the UN figure is based on a definition of forest as being an area with as little as 10% actual tree cover, which would therefore include areas that are actually savannah-like ecosystems and badly damaged forests." Other critics of the FAO data point out that they do not distinguish between forest types, and that they are based largely on reporting from forestry departments of individual countries, which do not take into account unofficial activities like illegal logging.

Despite these uncertainties, there is agreement that destruction of rainforests remains a significant environmental problem. Up to 90% of West Africa's coastal rainforests have disappeared since 1900. In South Asia, about 88% of the rainforests have been lost. Much of what remains of the world's rainforests is in the Amazon basin, where the Amazon Rainforest covers approximately 4 million square kilometres. The regions with the highest tropical deforestation rate between 2000 and 2005 were Central America—which lost 1.3% of its forests each year—and tropical Asia. In Central America, two-thirds of lowland tropical forests have been turned into pasture since 1950 and 40% of all the rainforests have been lost in the last 40 years. Brazil has lost 90–95% of its Mata Atlântica forest., Paraguay was losing its natural semi humid forests in the country's western regions at a rate of 15.000 hectares at a randomly studied 2 month period in 2010, Paraguay's parliament refused in 2009 to pass a law that would have stopped cutting of natural forests altogether.

Madagascar has lost 90% of its eastern rainforests. As of 2007, less than 1% of Haiti's forests remained. Mexico, India, the Philippines, Indonesia, Thailand, Burma, Malaysia, Bangladesh, China, Sri Lanka, Laos, Nigeria, the Democratic Republic of the Congo, Liberia, Guinea, Ghana and the Ivory Coast, have lost large areas of their rainforest. Several countries, notably Brazil, have declared their deforestation a national emergency. The

World Wildlife Fund's ecoregion project catalogues habitat types throughout the world, including habitat loss such as deforestation, showing for example that even in the rich forests of parts of Canada such as the Mid-Continental Canadian forests of the prairie provinces half of the forest cover has been lost or altered.

Controlling deforestation

Reducing Emissions

Main international organizations including the United Nations and the World Bank, have begun to develop programs aimed at curbing deforestation. The blanket term Reducing Emissions from Deforestation and Forest Degradation (REDD) describes these sorts of programs, which use direct monetary or other incentives to encourage developing countries to limit and/or roll back deforestation. Funding has been an issue, but at the UN Framework Convention on Climate Change (UNFCCC) Conference of the Parties-15 (COP-15) in Copenhagen in December 2009, an accord was reached with a collective commitment by developed countries for new and additional resources, including forestry and investments through international institutions, that will approach USD 30 billion for the period 2010–2012. Significant work is underway on tools for use in monitoring developing country adherence to their agreed REDD targets. These tools, which rely on remote forest monitoring using satellite imagery and other data sources, include the Center for Global Development's FORMA (Forest Monitoring for Action) initiative and the Group on Earth Observations' Forest Carbon Tracking Portal. Methodological guidance for forest monitoring was also emphasized at COP-15 The environmental organization Avoided Deforestation Partners leads the campaign for development of REDD through funding from the U.S. government.

In evaluating implications of overall emissions reductions, countries of greatest concern are those categorized as High Forest Cover with High Rates of Deforestation (HFHD) and Low Forest Cover with High Rates of Deforestation (LFHD). Afghanistan, Benin, Botswana, Burma, Burundi, Cameroon, Chad, Ecuador, El Salvador, Ethiopia, Ghana, Guatemala, Guinea, Haiti, Honduras, Indonesia, Liberia, Malawi, Mali, Mauritania, Mongolia, Namibia, Nepal, Nicaragua, Niger, Nigeria, Pakistan, Paraguay, Philippines, Senegal, Sierra Leone, Sri Lanka, Sudan, Togo, Uganda, United

Republic of Tanzania, Zimbabwe are listed as having Low Forest Cover with High Rates of Deforestation (LFHD). Brazil, Cambodia, Democratic Peoples Republic of Korea, Equatorial Guinea, Malaysia, Solomon Islands, Timor-Leste, Venezuela, Zambia are listed as High Forest Cover with High Rates of Deforestation (HFHD).

Farming

New methods are being developed to farm more intensively, such as high-yield hybrid crops, greenhouse, autonomous building gardens, and hydroponics. These methods are often dependent on chemical inputs to maintain necessary yields. In cyclic agriculture, cattle are grazed on farm land that is resting and rejuvenating. Cyclic agriculture actually increases the fertility of the soil. Intensive farming can also decrease soil nutrients by consuming at an accelerated rate the trace minerals needed for crop growth.The most promising approach, however, is the concept of food forests in permaculture, which consists of agroforestal systems carefully designed to mimic natural forests, with an emphasis on plant and animal species of interest for food, timber and other uses. These systems have low dependence on fossil fuels and agro-chemicals, are highly self-maintaining, highly productive, and with strong positive impact on soil and water quality, and biodiversity.

Monitoring Deforestation

There are multiple methods that are appropriate and reliable for reducing and monitoring deforestation. One method is the "visual interpretation of aerial photos or satellite imagery that is labor-intensive but does not require high-level training in computer image processing or extensive computational resources". Another method includes hot-spot analysis (that is, locations of rapid change) using expert opinion or coarse resolution satellite data to identify locations for detailed digital analysis with high resolution satellite images. Deforestation is typically assessed by quantifying the amount of area deforested, measured at the present time. From an environmental point of view, quantifying the damage and its possible consequences is a more important task, while conservation efforts are more focused on forested land protection and development of land-use alternatives to avoid continued deforestation. Deforestation rate and total area deforested, have been widely used for monitoring deforestation in many regions, including the Brazilian Amazon deforestation monitoring by INPE. Monitoring deforestation is a very complicated process, which becomes even more complicated with the increasing needs for resources.

Forest Management

Efforts to stop or slow deforestation have been attempted for many centuries because it has long been known that deforestation can cause environmental damage sufficient in some cases to cause societies to collapse. In Tonga, paramount rulers developed policies designed to prevent conflicts between short-term gains from converting forest to farmland and long-term problems forest loss would cause, while during the 17th and 18th centuries in Tokugawa, Japan, the shoguns developed a highly sophisticated system of long-term planning to stop and even reverse deforestation of the preceding centuries through substituting timber by other products and more efficient use of land that had been farmed for many centuries. In 16th century Germany landowners also developed silviculture to deal with the problem of deforestation. However, these policies tend to be limited to environments with good rainfall, no dry season and very young soils (through volcanism or glaciation). This is because on older and less fertile soils trees grow too slowly for silviculture to be economic, whilst in areas with a strong dry season there is always a risk of forest fires destroying a tree crop before it matures.

In the areas where "slash-and-burn" is practiced, switching to "slash-and-char" would prevent the rapid deforestation and subsequent degradation of soils. The biochar thus created, given back to the soil, is not only a durable carbon sequestration method, but it also is an extremely beneficial amendment to the soil. Mixed with biomass it brings the creation of terra preta, one of the richest soils on the planet and the only one known to regenerate itself.

Sustainable Practices

Certification, as provided by global certification systems such as Programme for the Endorsement of Forest Certification and Forest Stewardship Council, contributes to tackling deforestation by creating market demand for timber from sustainably managed forests. According to the United Nations Food and Agriculture Organization (FAO), "A major condition for the adoption of sustainable forest management is a demand for products that are produced sustainably and consumer willingness to pay for the higher costs entailed.

Certification represents a shift from regulatory approaches to market incentives to promote sustainable forest management. By promoting the positive attributes of forest products from sustainably managed forests, certification focuses on the demand side of environmental conservation." Rainforest Rescue argues that the standards of organizations like FSC are too closely connected to industry interests and therefore do not guarantee environmentally and socially responsible forest management. In reality, monitoring systems are inadequate and various cases of fraud have been documented worldwide.

Some nations have taken steps to help increase the amount of trees on Earth. In 1981, China created National Tree Planting Day Forest and forest coverage had now reached 16.55% of China's land mass, as against only 12% two decades ago

Using fuel from bamboo rather than wood results in cleaner burning, and since bamboo matures much faster than wood, deforestation is reduced as supply can be replenished faster.

Reforestation

In many parts of the world, especially in East Asian countries, reforestation and afforestation are increasing the area of forested lands. The amount of woodland has increased in 22 of the world's 50 most forested nations. Asia as a whole gained 1 million hectares of forest between 2000 and 2005. Tropical forest in El Salvador expanded more than 20% between 1992 and 2001. Based on these trends, one study projects that global forest will increase by 10%—an area the size of India—by 2050.

In the People's Republic of China, where large scale destruction of forests has occurred, the government has in the past required that every able-bodied citizen between the ages of 11 and 60 plant three to five trees per year or do the equivalent amount of work in other forest services. The government claims that at least 1 billion trees have been planted in China every year since 1982. This is no longer required today, but March 12 of every year in China is the Planting Holiday. Also, it has introduced the Green Wall of China project, which aims to halt the expansion of the Gobi desert through the planting of trees. However, due to the large percentage of trees dying off after planting (up to 75%), the project is not very successful. There has been a 47-million-hectare increase in forest area in China since the

1970s. The total number of trees amounted to be about 35 billion and 4.55% of China's land mass increased in forest coverage. The forest coverage was 12% two decades ago and now is 16.55%.

An ambitious proposal for China is the Aerially Delivered Reforestation and Erosion Control System and the proposed Sahara Forest Project coupled with the Seawater Greenhouse.

In Western countries, increasing consumer demand for wood products that have been produced and harvested in a sustainable manner is causing forest landowners and forest industries to become increasingly accountable for their forest management and timber harvesting practices.

The Arbor Day Foundation's Rain Forest Rescue program is a charity that helps to prevent deforestation. The charity uses donated money to buy up and preserve rainforest land before the lumber companies can buy it. The Arbor Day Foundation then protects the land from deforestation. This also locks in the way of life of the primitive tribes living on the forest land. Organizations such as Community Forestry International, Cool Earth, The Nature Conservancy, World Wide Fund for Nature, Conservation International, African Conservation Foundation and Greenpeace also focus on preserving forest habitats. Greenpeace in particular has also mapped out the forests that are still intact and published this information on the internet. World Resources Institute in turn has made a simpler thematic map showing the amount of forests present just before the age of man (8000 years ago) and the current (reduced) levels of forest. These maps mark the amount of afforestation required to repair the damage caused by people.

Forest Plantations

To meet the world's demand for wood, it has been suggested by forestry writers Botkins and Sedjo that high-yielding forest plantations are suitable. It has been calculated that plantations yielding 10 cubic meters per hectare annually could supply all the timber required for international trade on 5% of the world's existing forestland. By contrast, natural forests produce about 1–2 cubic meters per hectare; therefore, 5–10 times more forestland would be required to meet demand. Forester Chad Oliver has suggested a forest mosaic with high-yield forest lands interpersed with conservation land.

In the country of Senegal, on the western coast of Africa, a movement headed by youths has helped to plant over 6 million mangrove trees. The trees will protect local villages from storm damages and will provide a habitat for local wildlife. The project started in 2008, and already the Senegalese government has been asked to establish rules and regulations that would protect the new mangrove forests.

Afforestation and Biodiversity Conservation

Both afforestation and reforestation refer to the conversion of land under other uses to forest. Afforestation is defined as the direct human-induced conversion of land that has not been forested for a period of at least 50 years to forested land through planting, seeding, and/or the humaninduced promotion of natural seed sources. Reforestation is defined as the direct humaninduced conversion of non-forested land to forested land through planting, seeding, and/or the humaninduced promotion of natural seed sources on land that was forested but that has been converted to non-forested land.

Reforestation activities will be limited to reforestation occurring on those lands that, had been forested once, but that did not contain forest on 31 December 1989. The time limit included in the definitions is important; since only reforestation activities in areas that were non-forested prior to 1990 can be accounted for, it is thought that activities under the Kyoto Protocol do not generally create a perverse incentive for conversion of natural forests into plantation forests.

However, this incentive has not been totally removed as lands that did not contain forest as of 1990 but may have since been reforested, e.g. through natural forest succession, will be eligible for reforestation activities. These two activities are the only carbon uptake activities that are also eligible under the CDM. However, at this time, it is unclear whether the same definitions of reforestation and afforestation under the CDM will apply.

In planted forests, species selection often results in a trade-off between fast carbon assimilation and subsequent release vs. slower carbon assimilation and longer retention time. How these tradeoffs are made will affect biodiversity. This implies that fast rate of carbon uptake from the atmosphere and long retention time of the sequestered carbon cannot be maximised at the same time. In many types of tree plantations, soil carbon continues to be lost during the first 10-20 years due to continued leaching, and net accumulation only becomes positive with increased time the length of which is likely ecosystem-dependent.

The total carbon pool of a carbon-sequestering activity, the rate of positive change of the pool, and the time that carbon will remain sequestered in the system, strongly depend not only on climate, soil nutrients, and rotation length, but also on the dominant tree species.

For example in temperate forests, poplars (Populus) are fast-growing, may become very large, but are short-lived, while oaks (Quercus) and beeches (Fagus) are slowgrowing, also become very large, but are very longlived. Forests of the latter species are less ephemeral than poplar forests where in turn, dead wood is more rapidly decomposed.

From a biodiversity perspective, the choice of tree species can greatly affect the types of animals and associated understory plant species that can be supported. The use of either short- or longlived species depends on the goals. Longlived forest ecosystems, support more complex relationships than do simple and hence shorter-lived forests; therefore, the former support greater levels of biodiversity. A decision on how to balance the alternative goals for carbon and biodiversity will have to be made in any forest carbon-uptake activity.

Impact of Afforestation and Reforestation

Afforestation and reforestation projects can have positive, neutral or negative impacts on biodiversity. The impact depends on the level and nature of biodiversity of the ecosystem being replaced, the spatial scale being considered, and other spatial design and implementation issues.

Afforestation and reforestation activities may help to promote the return, survival, and expansion of native plant and animal populations. Degraded lands may offer the best opportunities for such activities, as these lands have already lost much of their original biodiversity. Plantations may allow the colonisation and establishment of diverse understory communities by providing shade and ameliorating harsh microclimates.

Specific sites may be better candidates for implementing such activities than others, based on past and present uses, the local or regional importance of their associated biological diversity and proximity to nearby, natural forests. In particular, the reduction of forest fragmentation, e.g. by careful design of native plantation establishment and/or forest regeneration strategies sites to give the most functionally connected forest landscape possible would have positive impacts on biodiversity, improving ecosystem resilience and allowing species migration in response to climate change.

Plantations of exotic species may only be capable of supporting low levels of local biodiversity at the stand level, but they could contribute to biodiversity conservation if appropriately situated within the broader landscape context; e.g. connecting areas of natural forest enabling for species migration and gene exchange. Activities that maintain a high ecosystem service value contribute to both carbon-uptake and forest biodiversity conservation. An important aspect is the extent to which activities take into account concerns of the local and indigenous communities in meeting the carbon credit priorities of investors.

Incorporation of what is 'valuable biodiversity' from the local community perspective helps to strike a balance between biodiversity and carbon uptake, and promote longterm protection of plantings. The stipulation in the Marrakesh Accords that CDM projects must contribute to the sustainable development of the host country and may best be achieved by the CBD ecosystem approach may encourage project planners to design activities that conserve and enhance biodiversity.

Afforestation and reforestation plantations can have beneficial environmental impacts, especially if modifications are incorporated. Although plantations typically have lower biodiversity than natural forests, in some cases they can reduce pressures on natural forests by serving as sources of forest products, thereby leaving greater areas of natural forests for biodiversity conservation and provision of environmental services.

Afforestation and reforestation activities may also re-establish critical ecological functions, such as erosion control within degraded watersheds, and corridors within a fragmented landscape. Further, in some countries success at supporting at least some native species in plantation forests has been achieved, by paying attention to structure, stem density, and species mixing. In some instances, plantation forests have been shown to maintain considerable numbers of local species. Even modest changes in project design have the potential to significantly benefit biodiversity in plantation forests.

Significant biodiversity benefits can be achieved by allowing a portion of the stand on a landscape to age past maturity, by reducing chemical and insect control, and avoiding locales where rare or vulnerable ecosystems and species are present at the time of site selection. Finally, mixedspecies plantations have more overall ecosystem service value and therefore are more likely to be retained by local communities for a longer time than single-species plantations. However, it must be noted that under climate change, there is considerable uncertainty associated with the permanence of benefits.

Afforestation and reforestation activities that replace native non-forest ecosystems with nonnative species, or with a single or few species of any origin, can negatively affect biodiversity. For example, in South Africa, expansion of commercial plantations has led to significant declines in several endemic and threatened species of native grassland birds and suppression of indigenous ground flora. Similarly, drainage of wetlands for afforestation and reforestation activities may not be a viable carbon mitigation option, as drainage will lead to immediate loss of carbon stocks and potential loss of biodiversity.

Afforestation with non-indigenous species may result in higher rates of water uptake than by existing vegetation and this could cause significant reductions in streamflow especially in ecosystems where water is limiting. These changes could have adverse effects on in-stream, riparian, wetland, and floodplain biodiversity. For example, the water yield from catchments in South Africa was significantly reduced when the catchments were planted with pines and eucalypts. Tree improvement through silvicultural techniques can increase the productivity associated with plantations, and maintain genetic diversity of local species.

Individual tree species are adapted to specific ranges of moisture and temperature. Careful selection of seeds and tree stocks under climate change scenarios, based on modelling, will enable more rapid growth and increase survivorship of planted tree species and individuals than would be expected by relying on available stocks. This can be accomplished by matching expected temperature and moisture regimes to planted species and individuals within species and paying attention to maintaining species genetic diversity to enhance success of plantation forests.

Conversely, single-species plantations of commercially valuable tree species have been widely planted in many regions of the world. While still within their geographic range, these plantations have often been planted off-site into areas where factors like soil, elevation, moisture, slope, and aspect differ significantly from where they are normally found in the landscape. Many of these plantations will become susceptible to reduce growth or dieback under drier or warmer climate scenarios.

Measuring the success of afforestation and reforestation activities can be accomplished with a series of indicators for carbon uptake, as well as for biodiversity, at the site and landscape scale. In developing such activities, the following considerations for biodiversity may be useful:

— landscape structure and planted trees species composition can affect understory plant species and animal species diversity;

— a regional suite of animal species requires the full variety of local forest types and ages of stands, with the structures normally associated with those forests;

— planted forests that are structurally diverse maintain more species than those that have simple structure;

— planted forests of native species conserve local and regional animal species better than do plantations of exotic tree species, or monocultures of native species;

— large areas of forest maintain more species than do small areas, and fragmented forests maintain fewer species than do continuous forests;

— core areas and protected areas connected by reforested corridors or habitats enhance population levels of species by reducing fragmentation effects and improving dispersal capability, and through supporting more individuals;

— some exotic tree species have the potential to become invasive, with potentially negative consequences for ecosystem functioning and biodiversity conservation;

— planted forests that have high genetic diversity are likely to be more successful over time and under climate changes than those with reduced genetic diversity.

— the spatial context where activities take place is important to optimise for biodiversity of desired species.

Uncertainty pertaining to the benefits of mitigation and adaptation measures suggests that adaptive management should be designed into any project. Afforestation and reforestation projects should be viewed as experiments with respect to their possible benefits to biodiversity. Monitoring programmes should be put in place to enable the longterm assessment of benefits compared to expectations, and possible adjustments made as required to design and future efforts.

Mires and Peatlands

Pristine mires play an important role with respect to global warming as carbon stores. Their impact on climate change due to the emission of methane (CH_4) and nitrous oxide (N_2O) is typically insignificant. However, methane production can be high when water tables are within 20 cm of the surface. Mires and peatlands are characterised by their unique ability to accumulate and store dead plant material originating from mosses, sedges, reeds, shrubs, and trees (i.e., peat), under waterlogged conditions. About 50% of the dry organic matter of peat consists of carbon.

Peatlands are the most prevalent wetland in the world, representing 50 to 70 percent of all wetlands and covering more than four million km^2 - or three percent - of the land and freshwater surface of the planet. Between 270-370 Gt of carbon is currently stored in the peats of boreal and sub-boreal peatlands alone. This means that, globally, peat represents about one-third of the total soil carbon pool. Peat contains the equivalent of approximately 2/3 of all carbon in the atmosphere and carbon equivalent to all terrestrial biomass on the earth. Peatlands exist on all continents, from tropical to polar zones, and from sea level to high altitude.

Humans affect peatlands both directly, through drainage, land conversion, excavation, and inundation, and indirectly, as a result of air pollution, water contamination, water removal, and infrastructure development. Anthropogenic drainage has changed mires and peatlands from a global carbon sink to a global carbon source, and afforestation and reforestation activities in recently drained peatlands may be inconsequential as carbon sequestration activities.

Human activities continue to be the most important factors affecting peatlands, both globally and locally, leading to a current annual decrease of the mire resource. When peatlands are drained to create more agricultural land N_2O emissions are increased and these lands become more prone to fires. In some years greenhouse gas emissions from the burning of these drained peatlands may constitute a substantial portion of the global emissions.

Agroforestry Systems

Agroforestry systems incorporate trees or shrubs in agricultural landscapes. Agroforestry practices could be considered eligible under the CDM if they meet the adopted CDM definition of afforestation or reforestation. Agroforestry systems include a wide variety of practices: agrosilvicultural systems; silvopastoral systems; and tree-based systems such as fodder plantations, shelterbelts, and riparian forest buffers. These systems are typically managed, but can also be natural, such as silvopastoral systems in Sudan. Agroforestry systems may lead to more diversified and sustainable production systems than farming systems without trees, and may provide increased social, economic and environmental benefits.

The IPCC recognises two classes of agroforestry activities for increasing carbon stocks:

(a) land conversion; and

(b) improved land use.

Land conversion includes transformation of degraded cropland and grassland, into new agroforests. Improved land use requires the implementation of practices such as high-density plantings and nutrient management that result in increased carbon stock. Globally, significant amounts of carbon could be sequestered in agroforestry systems, due to the large agricultural land base in many countries. In temperate systems, agroforestry practices have been shown to store large amounts of carbon in trees and shrubs.

Positive net differences in carbon stocks, including those in the soil, have been documented in the tropics between agroforestry systems and common agricultural practices. In addition to carbon uptake, agroforestry activities can have beneficial effects on biodiversity, especially in landscapes that are dominated by production agriculture. Agroforestry can add plant and animal diversity to landscapes that might otherwise contain only monocultures of crops. The important role of farmland habitat for the conservation of native plant species in Eastern Canada. In the Great Plains region of the United States, where cropland occupies most of the landscape, linear riparian zones and field shelterbelts play essential roles in maintaining natural habitats for biodiversity.

In the same region, Brandle et al highlighted the potential of agroforestry practices to provide wildlife habitat. Traditional agroforestry systems, e.g. shaded coffee plantations, are common throughout Central and South America. These systems may contain well over 100 annual and perennial plant species per field and provide beneficial habitat for birds and other vertebrates. Agroforestry can enhance biodiversity on degraded and deforested sites. Agroforestry systems tend to be more biologically diverse than conventional croplands, degraded grasslands or pastures, and the early stages of secondary forest fallows. However, where agroforestry replaces native forests biodiversity is usually lost.

The use of native species in agroforestry systems will provide greatest benefits to biodiversity. In view of human migrations to the forest margins, the optimal tradeoffs between carbon sequestration and economic and social benefits are an important policy determination. Agroforestry can be used to functionally link forest fragments and other critical habitat as part of a broad landscape management strategy. Agroforestry can augment the supply of forest habitat and enhance its connectivity. This can facilitate the migration of species in response to climate change. Even when there are forest reserves in an area, they may be too small in size to contain the habitat requirements of all animal species, and whose populations may extend in range beyond reserve boundaries.

Effects of Revegetation on Biodiversity

Revegetation is defined as a direct humaninduced activity to increase on-site carbon stocks through the establishment of vegetation that covers a minimum area of 0.05 hectares and does not meet the definitions of afforestation and reforestation. Revegetation includes various activities designed to increase plant cover on eroded, severely degraded or otherwise disturbed land.

Short-term goals of revegetation are often erosion control, improved soil stability, recovery of soil microbial populations, increased productivity of degraded rangelands and improved appearance of sites damaged by such activities as mining or construction. It is often the initial step in the longterm restoration of ecosystem structure and function, natural habitats, and ecosystem services.

Soils of eroded or degraded sites generally have low carbon levels but have high potential for carbon sequestration through revegetation. The sequestration potential of eroded land restoration as 0.2-0.3 Gt of carbon yr^{-1}. Research in Iceland has demonstrated sequestration of carbon in soils, above- and below-ground biomass, and litter, but sequestration rates depend on various factors, including the revegetation method, soil characteristics, and climate.

The effects of revegetation on biodiversity will vary depending on the site conditions and methods used. Effects on biodiversity can be positive if revegetation efforts create conditions that are conducive to an increase of native plant species over time, or if it prevents further degradation and protects neighbouring ecosystems. Conversely, biodiversity can be negatively affected by revegetation if it results in conditions that impede the colonisation of native species. In certain instances where endemic species may now be impossible to grow on some severely degraded sites, the use of exotic species and fertilisers may provide the best opportunity as a catalyst for regeneration of natural vegetation.

However, in such instances, it is desirable that the use of exotic species is temporary. Furthermore, exotic species used for revegetation can invade native habitats and alter plant communities and ecosystem processes far beyond the areas where they were originally used. Revegetation actions that do not depend on direct seeding or planting enhance local populations and have positive effects on biodiversity. Such actions involve manipulation of: seed dispersal processes, seedbed properties and resource base for establishment and growth of plants. This should enhance local populations and have positive effects on biodiversity, unless exotic species are common at the given site.

Land Management Actions

Land management actions to offset greenhouse gas emissions can affect overall environmental quality, including soil quality and soil erosion, water quality, air quality, and wildlife habitat and in turn, affect terrestrial and aquatic biodiversity. The subsections below deal with management of forests, croplands and grazing lands.

Management of Forest

Most of the world forests are managed, so improved management can enhance carbon uptake, or at least minimise carbon losses, and maintain biodiversity. For the purposes of the Kyoto Protocol, forest management is defined as a system of practices for stewardship and use of forest lands, aimed at fulfilling relevant ecological, economic, and social functions of the forest in a sustainable manner.

Forest management refers to activities such as harvesting, thinning and regeneration. These management activities provide opportunities to promote conditions that are conducive to increased biodiversity. Improved forest management in central Africa could provide the uptake of an additional 18.3 Gt of carbon over the next 50 years. By reducing the amount of logging debris through "good" forestry practices such as low-impact harvesting in tropical forests, significant amounts of carbon in the standing vegetation can be retained and that otherwise would have been released to the atmosphere by decomposition.

Low impact harvesting also minimises the probability of forest fires, as there is little woody debris that otherwise would serve as combustion fuel. Forest ecosystems are extremely varied and the positive or negative impacts on biodiversity of any forest management operation will differ according to soil, climate, and site history. Therefore, it would not be helpful to recommend that any specific system or measure is inherently good or bad for biodiversity under all circumstances.

Prescriptions must be adapted to specific local forest conditions and the type of forest ecosystem under management. Because forests are enormous repositories of terrestrial biodiversity at all levels of organisation, good management practices can have positive effects on biodiversity. Forestry practices that enhance biodiversity in managed stands and have a positive influence on carbon retention within forests include: increasing rotation age, low intensity harvesting, leaving woody debris, post-harvest silviculture to restore native plant communities that are similar to natural species composition, and harvesting that emulates natural disturbance regimes.

The application of appropriate silvicultural practices can reduce local impacts while ensuring the longterm protection of soils and animal and plant species. The use of appropriate harvesting methods can lessen the negative impacts on biodiversity, while still providing socioeconomic benefits to local owners and communities that are largely dependent on the forest for their livelihoods. Measuring progress towards sustainability, and managing adaptively, is an important aspect of forest management.

Many national and international agencies, have adopted a series of indicators to measure progress in conserving biological diversity in sustainable forest management, for which there is a large body of available literature. Forest regeneration includes practices such as planting at specific stocking levels, enrichment planting, reduced grasing of forested savannas, and changes in tree provenances/genetics or tree species. Regeneration techniques can influence species composition, stocking, and density and can affect biodiversity.

Natural regeneration of forests can provide benefits for biodiversity by expanding the range of natural or semi-natural forests. Areas adjacent to natural forests demonstrate the most potential for such activities. Plantations, even of indigenous species, adjacent to natural or semi-natural forests may not provide maximum benefits to biodiversity unless designed as part of an integrated scheme for the eventual restoration of natural forests.

Efforts to understand and integrate land use at the landscape scale can increase the likelihood that biodiversity will be accommodated. Forest fertilisation may have negative or positive environmental effects. Fertilisation may adversely affect biodiversity, soil and water quality by improving the environment for unwanted species, by altering species composition and by increasing nutrients in water runoff that adversely affect watercourses. Although careful attention to the rate, timing, and method of fertilisation can minimise environmental impacts, in general, positive environmental benefits are not likely to result from forest fertilisation, except on highly degraded sites.

There, fertilisation may be necessary where soils and nutrients have been depleted. Fertility can affect the establishment of trees, shrubs, and understory plant communities. When organic and inorganic amendments were applied to an eroded site in South Iceland, plant cover and the diversity of native vascular plant and moss species increased. Another study on the same site showed increased carbon stocks in soil, vegetation and litter in similar but successively older treatments. Approaches that use not only inorganic fertilisers, but also organic amendments and nitrogen fixing plant species as well should be considered.

Fire management has environmental impacts that are difficult to generalise because in some forest ecosystems fires are essential for regenerative processes to occur. Restoring nearhistorical fire regimes may be an important component of sustainable forestry but may also require practices, such as road construction, that may create indirect deleterious environmental effects. The suppression of natural fire cycles leads to the excessive accumulation of combustible material, potentially leading to larger, more intense fires and is unlikely to provide viable longterm carbon sequestration.

In some forest ecosystems periodic fires are necessary to regenerate understory plant communities and their associated biodiversity. However, in forests not subjected to recurrent natural fires, e.g., tropical rainforests, increased fire frequencies lead to overall negative effects on biodiversity, and loss of soil nutrients through leaching and runoff. The use of biocides to control pests may result in increased or reduced biodiversity. Many introduced plant and animal species have had unintended negative impacts on biodiversity.

Carefully targeted pest management efforts have been used to reduce the impact of introduced species on native populations, for example predation of birds and their eggs. Biocides can, at times, prevent large-scale forest die-off, and can increase benefits associated with landscape, recreation, and watersheds. Conversely, the potential adverse effects of herbicides and pesticides on biodiversity include disruption of root-mycorrhisae symbiosis and a reduction in plant species populations and diversity.

Pesticide use may also have undesired secondary effects on predators. If not carefully used, pesticides can be leached into surface waters and groundwater and cause negative impacts to aquatic biodiversity and human health. Harvesting practices affect the quality and quantity of timber produced, which has implications for carbon storage and biodiversity. Harvesting can have positive or negative impacts on biodiversity, recreation, and landscape management. Small-scale harvesting is often appropriate in forest ecosystems on soils that are subject to erosion.

Cropland Management

The Marrakesh Accords define cropland management as "the system of practices on land on which agricultural crops are grown, and on land that is set aside or temporarily not being used for crop production". Most carbon stocks in cropland are housed in the soil; they currently constitute about 8-10 percent of total global carbon stocks.

Some studies suggest that most of the world's agricultural soils have about half of their pre-cropped soil carbon and that change in soil management, especially reducing tillage, can greatly increase their carbon stocks. Generally, conversion of natural systems to cropland results in losses of soil organic carbon ranging from 20-50 percent of the pre-cultivation carbon stocks.

For example, following conversion from forest to rowcrop agriculture, soil carbon losses associated with CO_2 emissions is about 20-30 percent of the original carbon stocks. On a global basis, the cumulative historic loss of carbon from agricultural soils due to practices such as crop residue removal, inadequate erosion control and excessive soil disturbance has been estimated at 55 Gt, or nearly one third of the total carbon loss from soils and vegetation. Activities in the agricultural sector that reduce greenhouse gas emissions and increase carbon sequestration may enhance or decrease given levels of biodiversity.

There are many agricultural management activities that can be used to sequester carbon in soils. Practices may have positive or negative effects on biodiversity, depending on the specific practice and the context in which it is applied. Activities include adopting farmer participatory approaches; consideration of local knowledge and technologies; the use of organic materials; and the use of locally adapted crop varieties and crop diversification. Agricultural practices that enhance and preserve soil organic carbon can affect CH_4 and N_2O emissions.

Agricultural intensification practices that may enhance production and increase plant residue in soil include crop rotations, reduced bare fallow, cover crops, improved varieties, integrated pest management, optimisation of inorganic and/or organic fertilisation, irrigation, water table management, and site-specific management. These have numerous ancillary benefits including increased food production, erosion control, water conservation, improved water quality, and reduced siltation of reservoirs and waterways benefiting fisheries and biodiversity.

However, soil and water quality is adversely affected by indiscriminate use of chemical inputs and irrigation, and increased use of nitrogen fertilisers will increase fossil energy use and may increase N_2O emissions. Agricultural intensification influences soil carbon through the amount and quality of carbon returned to the soil, and through water and nutrient influences on decomposition. Irrigation can increase crop production, but may also degrade ecosystems.

Irrigation also increases the risk of salinisation and may divert water from rivers and flood flows with significant impacts on the biodiversity of rivers and flood plains. Return flows from irrigation can cause downstream impacts on water quality and aquatic ecosystems. Additional impacts can include the spread of water-borne diseases. Conservation tillage denotes a wide range of tillage practices, including chisel-plow, ridge-till, strip-till, mulch-till, and no-till to conserve soil organic carbon.

Adoption of conservation tillage has numerous ancillary benefits, including control of water and wind erosion, water conservation, increased water-holding capacity, reduced compaction, improved soil, water, and air quality, enhanced soil biodiversity, reduced energy use, reduced siltation of reservoirs and waterways with associated benefits for fisheries and biodiversity. In some areas, increased leaching from greater water retention with conservation tillage could cause downslope salinisation.

Reduction or elimination of intensive soil tillage practices can preserve and increase soil organic carbon stocks. In these practices 30% or more of crop residues are left on the soil surface after planting. Conservation tillage has the potential to sequester significant amounts of carbon in the soil. Soil carbon sequestration can be further increased when cover crops are used in combination with conservation tillage.

Carbon levels can be increased in the soil profile for 25 to 50 years, or until saturation is reached, but the rate may be highest in the initial 5 - 20 years. However, longterm soil carbon sequestration through conservation tillage will largely depend on its continued use, as reversion back to conventional practices can cause the rapid loss of sequestered carbon. Erosion control practices-which include water conservation structures, vegetative strips used as filters for riparian zone management, and agroforestry shelterbelts for wind erosion control-can reduce the global quantity of soil organic carbon displaced by soil erosion.

There are numerous ancillary benefits and associated impacts, including increased productivity; improved water quality; reduced use of fertilisers, especially nitrates; decreased siltation of waterways; reduced CH_4 emissions; associated reductions in risks of flooding; and increased biodiversity in aquatic systems, shelter belts, and riparian zones. Rice management strategies-which include irrigation, fertilisation, and crop residue management-affect CH_4 emissions and carbon stocks. But there is limited information on the impacts of greenhouse gas mitigation rice management activities on biodiversity.

Management of Grasing Lands

The response of grasing land systems will vary under potential climate change scenarios depending on its type and location. Grasing land (which include grasslands, pasture, rangeland, shrubland, savanna, and arid grasslands) contain 10-30 percent of the world's soil carbon. The mixture of grass, herb, trees and shrub species usually determines the productivity of a given rangeland. Grasing land with a higher percentage of grass in relation to other plant components is likely to have higher productivity.

A greater percentage of annual or ephemeral species would suggest lower annual productivity, whereas a predominance of perennial species is more likely to result in high productivity. The Marrakesh Accords define grasing land management as "the system of practices on land use for livestock production aimed at manipulating the amount and type of vegetation and livestock produced". Operationally, a distinction is sometimes made between grasing land management and grassland management; grasing lands are managed for livestock, whereas grasslands may be managed for different purposes, including conservation, but not specifically for livestock.

One of the goals of grasing land management is to prevent overgrasing, which is the single greatest cause of grassland degradation and the overriding human-influenced factor in grassland soil carbon loss. In grasing lands, carbon accumulates above- and below-ground, and transforming cropped or degraded lands to perennial grasslands can increase above- and below-ground biomass, soil carbon, and biodiversity.

Protection of previously intensively grased lands and reversion of cultivated lands to perennial grasslands is likely to be more prevalent in countries with agricultural surpluses, but opportunities for environmental protection set-asides are possible in all countries. Globally, estimates of the potential area of cropland that could be placed into set-asides are approximately 100 M ha. Grasslands management activities that can be used to sequester carbon in soils include grasing management, protected grasslands and set-asides, grassland productivity improvements, and fire management.

The productivity of many pastoral lands, and thus the potential for carbon sequestration, particularly in the tropics and arid zones, is restricted by nitrogen and other nutrient limitations and the unsuitability of some native species to high-intensity grasing. Introduction of nitrogen-fixing legumes and high-productivity grasses or additions of fertiliser can increase biomass production and soil carbon pools, but some of these introduced species have significant potential to become weeds.

Most grassland management activities are beneficial to biodiversity and carbon uptake; some such as fertilisation may decrease on-site biodiversity. Carbon accumulation can be enhanced through improved practices when grasing lands are intensively managed or strictly protected. Properly managed native species can enhance the biodiversity associated with grasing lands.

Native species are also often more tolerant of climatic variations than exotic species, and can provide essential habitat for animals. Native perennial species of grass have the potential to establish and effectively compete with annuals, improving system stability.

References

Canadell, J.G.; M.R. Raupach (2008-06-13). "Managing Forests for Climate Change". *Science*(AAAS) 320 (5882): 1456–1457.

Whitney, Gordon G. (1996). *From Coastal Wilderness to Fruited Plain : A History of Environmental Change in Temperate North America from 1500 to the Present.* Cambridge University Press.

Williams, Michael. (2003). *Deforesting the Earth.* University of Chicago Press, Chicago.

Wood well, G.M.; North, WJ (1988-12-16). "CO2Reduction and Reforestation". *Science* (ALAS) 242(4885): 1493–1494.

Wunder, Sven. (2000). *The Economics of Deforestation: The Example of Ecuador.* Macmillan Press, London.

4

Environmental Pollution and Biodiversity

Concerns about the environmental effects of air pollution stretch back for hundreds of years. In 1661, the English pamphleteer John Evelyn wrote Fumifugium - or the smoake of London dissipated Evelyn 1661, sic) about air pollution in the capital, and the term "acid rain" was first used in the mid 19th century in the north of England (Smith, 1872). Maps of sulphur dioxide levels drawn using the decline of lichens as the system of measurement were available prior to the First World War. Ecological effects can be measured back for hundreds of years.

More recent interest in the long-range effects of air pollution date from the 1960s. Attempts to control local air pollution problems, mainly by dispersal via high chimneys, resulted in the incorporation of sulphur and nitrogen dioxides into the atmosphere and the creation of sulphuric and nitric acids in the air. These fall to earth, sometimes hundreds of miles from their source, in the form of rain, mists and snow. A growing understanding about the ecological implications of so-called "acid rain" helped focus attention onto the issue of air pollution, and several other problems or potential problems were identified.

Pollutants

Acid rain is a general and simplified term used to describe a range of pollution effects. Several air pollutants can cause acidification of the environment. These include sulphur and nitrogen oxides (SO_2 and NO_x), which are given off when fossil fuels are burnt in power stations, industrial boilers and motor vehicles, and when plant material such as wood is burnt. Acidification occurs in two ways:

— either the gases convert chemically in the atmosphere, turning into acids and falling as rain, mists or snow;

— or they fall to earth as dry gases and are converted to acids through the action of rainwater.

These pollutants can also cause ecological damage in their gaseous form. Other important gaseous air pollutants occur, including hydrocarbons, which are pollutants themselves and can also react with nitrogen oxides in the presence of sunlight to form photochemical ozone (O_3), itself an important pollutant in the troposphere. To a lesser extent, ammonia (NH_4) from livestock slurry, and trace metals from industrial processes, also have important effects on the environment.

Critical Loads

Some measure of the importance of these effects, from an ecological perspective, can be gained from the use of the critical load concept.

A critical load is the quantitative estimate of an exposure to one or more pollutants below which significant harmful effects on sensitive elements of the environment do not occur according to present knowledge, ie a measure of the damage threshold for pollutants. Critical loads can be set for a range of different habitats and species. Scientists acting under the auspices of the United Nations Economic Commission for Europe (UNECE) have collated critical load data for sulphur and acidity levels throughout Europe, and have produced maps showing where the tolerance of soils and waters is already exceeded, or is likely to be exceeded in the future. A recent research project for WWF pinpoints important European nature conservation areas that are likely to be at high risk from air pollution. Under controls proposed by the 1985 sulphur protocol, some 71 per cent of the protected areas studied are in areas suffering excess acid pollution. Even if countries were to adopt far more radical environmental scenarios, between 20-25 per cent of Europe's protected areas would remain at risk from acidification.

Air Pollution and Biodiversity

Several attempts have been made to analyze the impacts of air pollution on wildlife. More recently, research for WWF has assessed the impacts on wildlife through a literature survey which identified effects on 1,300 species, including 11 mammals, 29 birds, 10 amphibians, 398 higher plants, 305 fungi, 238 lichens and 65 invertebrates, providing the most detailed survey to date. In general, the studies have concentrated on either specific ecosystems, or individual groups of plants and animals.

Whilst these investigations have all been useful in helping to identify the existence and scale of the problem relating to biodiversity and air pollution, they have not, on the whole, attempted to look at general trends. Drawing on the overviews referred to above, and on other published papers, the current paper proposes some general ecological considerations regarding the issue, and backs these with relevant data and examples.

General Considerations

Lower life forms are usually more affected by air pollution than higher life-forms.

Early attempts to look at the link between air pollution and wildlife focused mainly on the so-called "charismatic megafauna", ie on large and "colourful" species of animals. In fact, the most widely affected species - in terms of both number of species suffering damage from air pollution and also sensitivity of individual species to pollution - are amongst the lower life forms. In particular, lichens, bryophytes, fungi, and soft-bodied aquatic invertebrates are likely to be at risk.

Impacts of pollution in these high risk groups are likely to be general across many species, and directly related to the toxic effects of pollution itself. On the other hand, impacts on higher plants and, particularly, on higher animals are likely to be limited to sensitive species, and to act on the whole through secondary affects, such as changes to food supply, or inter-specific competition.

For example, both gaseous sulphur dioxide pollution and acid deposition are known to damage literally hundreds of lichen species in the UK. Air pollution has caused the extirpation of many species from industrial areas and the decline of others, even in remote parts of western Britain1 On the other hand, years of research have to date only found two birds whose range has been affected; the house martin (Delichon urbica) by sulphur dioxide and the dipper (Cinclus cinclus) by the impacts of freshwater acidification on its food species (although there may also have been some impacts on fish feeding birds). Neither of these species appears at risk of serious decline, and the former has now recolonised some areas due to a decline in SO_2

In general, plants are more affected by air pollution than animals on land, but not in freshwater.

Although precise comparative studies have not been carried out, there seem to be greater losses amongst terrestrial plant communities than amongst land animals under conditions of high air pollution. By their nature, plants are less able to adapt to sudden changes in pollution levels and climatology than animals, which often have the option of moving or changing food source. For example, in the literature survey referred to above1, evidence was found for pollution effects on over three times as many terrestrial plants as animals. Whilst some of these differences may indicate a bias towards certain groups amongst researchers, it accords well with other findings referred to above.

This situation apparently changes in freshwater ecosystems, where decline due to increasing acidity is greater among animals than plants. Studies of benthic fauna in Sweden found that diversity amongst animal species declined by 40 per cent for a pH reduction of 1 unit, while plant species declined by only 25 per cent under the same conditions.

Studies suggest that if a species is affected by air pollution at all, it is likely to decline. However, a minority of species thrive under polluted conditions. There are two reasons for this:

— some species appear to be stimulated by pollutants. For example, many aphids grow faster in conditions of high sulphur dioxide and nitrogen oxides;
— some species are resistant to pollution and expand to fill the spaces left by the disappearance of more sensitive species.

Impacts on Wild Plants and Animals

Most studies of wildlife effects have concentrated on individual species or particular groups. In the following section, an attempt is made to synthesise this information into a more general analysis of impacts.

Air pollution has played a key role in changing the distribution of plant species and the ecology of susceptible plant communities in polluted regions

Air pollution affects plants in many ways which have implications for overall biodiversity and ecology. Effects have been studied in detail for lichens1 and trees and also researched for bryophytes, fungi and herbaceous flowering plants. It is clear that susceptible individuals in all these groups can be affected by pollution, although debate remains in some cases about both the severity and the threshold of effects. Impacts occur as a result of various factors, including:

— Direct toxic effects on adult plants from either gaseous pollutants or acid deposition: these effects have been studied in particular detail for some crop species, but results remain relevant for many wild species as well. Interaction between gaseous and wet acid deposition also sometimes changes the nature of the

— Toxic effects on plants' reproductive capacity: there is evidence that air pollution can reduce some plants' ability to reproduce, thus causing long-term changes to population

— Changes in soil fertility due to pollutant deposition, particularly of nitrogen compounds: increased deposition of nitrogen can sometimes have a fertilizing effect on plants, particularly in ecosystems where nitrogen levels are the factor controlling growth rate of plants. In other cases, an excess of nitrogen can, conversely, reduce growth

— Changes in soil acidification: airborne acid pollution has been linked to accelerated acidification of soil in base-poor environments, and to a consequent decline in calcicole (calcium-loving) plants, potential aluminium toxicity, leaching of nutrients and base cations, effects on mycorrhizae etc.

— Increased or decreased competition from other plants: in polluted ecosystems, a small number of resistant plant species can dominate plant communities. For example, green algae such as Pleurococcus vulgaris can replace epiphytic lichens on trees, while Spahgnum species regularly replace other macrophytes in acidified waters.

— Increased predation through impacts of air pollutants on plant pests such as aphids: growth in many aphid species is increased by exposure to atmospheric sulphur dioxide and nitrogen oxides, and also in some cases to mixtures of pollutants. There is now strong evidence that aphid predators will not be able to keep up with this population increase and that the health of feed plants will suffer in consequence.

The end results include changes in the structure of plant communities. After initial research that concentrated mainly on commercially-valuable trees, crop plants and lichens, evidence has now also accumulated on effects on other wild plants.

Most evidence linking groups or species of invertebrates with population changes due to air pollution is statistical, ie few studies have been carried out into the mechanisms by which pollution is affecting invertebrate life cycles.

Data for land invertebrates tends to be more patchy, and based largely on statistical or circumstantial evidence. Whilst there is an increasing acceptance among scientists that some invertebrate species are damaged by air pollution, no large scale or general studies have been attempted, except on a limited scale for land molluscs.

Relatively few examples are known of higher animals suffering direct toxic effects from either acidity or gaseous air pollution. A number of mammals are known to build up high levels of heavy metals and other pollutants in contaminated environments. For example, cadmium levels in the internal organs of game animals in Sweden have prompted authorities to recommend that the kidneys of older elk are not eaten and that liver from game is not eaten more than once or twice a month. Deterioration in the antlers of roe deer (Capreolus capreolus) in Poland has been linked to sulphur and heavy metal pollution. Research in former Czechoslovakia found high sulphur levels in hares (Lepus capensis) living in polluted areas. Wild mink (Mustela vision) and Canadian otter (Lutra canadensis) have both been found to have high mercury levels near industrial sites. However, the long term ecological effects of these contamination levels remain unknown.

Measurable effects on wild animals, when they do occur, are generally due to either loss of food or loss of ability to reproduce. For example, studies on mammals and birds have found the strongest links between declines and loss of food species, often through freshwater acidification.

Amongst the animals of slightly lower orders, including particularly amphibians and fish, impacts are more commonly related to loss of reproductive capacity. In most cases, acidity itself does not appear to be the problem, but rather the impact that acidification has of releasing metals such as aluminium into the water. There has, in addition, long been a debate about the role that acidification and aluminium could play in eggshell thinning in certain bird species.

Apart from a few specialised feeders referred to above, most animals at or near the top of the food chain have proved adaptable to changing conditions created by atmospheric pollution. For example, unlike the dipper, the grey wagtail (Motacilla cinera) proved able to survive in acidified streams in Wales and Sweden, probably by changing its feeding from freshwater to bankside invertebrates. Pelagic pursuit feeding water birds such as divers (Gavia spp.) and the goosander (Mergus serrator) can compensate for reduced fish density in partly acidified lakes through better hunting success because of increased water transparency, due to disappearance of many algae. In a survey of 45 oligotrophic lakes in Sweden, goosanders and black throated divers (Gavia arctica) were found to favour partly acidified lakes. Adult divers appear capable of switching food for young from small fish to aquatic invertebrates, and in Sweden higher production of young occurred on acidified lakes, perhaps partly because of reduced predation from pike (Esox lucius).

These adaptations have their limits, and evidence from the USA suggests that if most or all the fish disappear from acidified lakes, divers (known as loons in North America) will decline. However, the fact that high or top predators can often adapt quite effectively to changing conditions means that their status under acidified or polluted conditions remains complex.

Responses to air pollution also differ markedly within many animal groups. These sometimes divide clearly between different subgroups, in other cases susceptibility or resistance to air pollution appears to be more individual. Some examples are given below:

— Terrestrial insects: distinct types of response to SO_2 pollution have been identified which distinguish some groups of land-living insect, for example:
  — Very susceptible: eg many butterflies and moths;
  — Moderately susceptible, eg the beetle Ips dentatus and the flatbug Aradus cinnamoneus;
  — Very tolerant and sometimes benefitted by SO_2 pollution: aphids.
— Stonefly larvae: Plectoptera: most species decline rapidly in acidified waters but a few, such as Amphinemura sulcicollis and Chloroperla tormentium can withstand high levels of acidity and in consequence will dominate acidified streams.

— *Fish:* variations in susceptibility to acidification occur both within and between species.

Complexities of Air Pollution

The previous section has given some indications of the scale and breadth of impacts on individual species. However, air pollution is far from a single or simple phenomenon. In the following section, some of the interactions between different pollutants, and between pollutants and other factors, are briefly examined.

Wild plants and animals do not face a single problem, or a simple range of pollution effects. The cocktail of atmospheric pollutants facing species in many parts of the world varies enormously, and each combination has a slightly different effect. Combinations can sometimes produce a joint effect greater than the sum of individual effects (synergism) and on other occasions effectively cancel each other out. Identifying a response, or a suspected response, to a mixture of pollutants is often easier than identifying the particular role that individual pollutants play in any observed responses, or discovering how the pollutant acts to cause changes. Our knowledge of pollutant interactions remains limited, but some information on varying responses has been built up over the last few years. For example:

— Some lichens are more sensitive to gaseous sulphur dioxide than to wet acid deposition, while in other species the reverse is true.

— Several Sphagnum moss species decline under conditions of high sulphur dioxide pollution and a few are also susceptible to nitrogen oxides. However, many of the same species increase in acidified waters, where wet acid deposition has reduced pH levels and eliminated other macrophyte plants.

— Fumigation experiments with crop plants have found a wide range of responses according to whether the plant is exposed to sulphur dioxide, nitrogen oxides, ozone or varying combinations and mixes of these and other pollutants.

Some pollutants can appear to be initially beneficial to particular species but later become harmful, or are harmful to the ecosystem as a whole.

Air pollution can benefit certain species at the expense of others, either because they are particularly resistant, or because the surrounding habitat changes in a way that benefits them over other species. For example:

— *Flowering plants:* Whilst soil acidification often leads to an overall loss in flowering plant variety, some species will expand as a result of increased nitrogen availability, lack of competition etc. Studies in Sweden found, for example, that dogs mercury (Mercuralis perennis), woodruff (Galium odoratum) and wood sorrel (Oxalis acetosella) all increased under conditions of acidification.

— *Insects*: At least twenty species of aphids show increased mean rate of growth under conditions of high levels of SO_2, NO_x or mixtures of the two. Experiments suggest that changes are mediated via the food plant in response to pollutant-induced changes in the plant. Increased growth rate is usually accompanied by increased reproduction. Whilst this boosts populations of aphids it also, in consequence, increases pressure on host plants and disrupts ecosystem stability.

— *Amphibians:* Research on the impacts of acidification on the survival of common frogs (Rana temporaria) suggests that early mortality of a proportion of eggs can actually increase the number surviving to adulthood in some cases, because of reduced competition and increased availability of food. However, this early mortality also reduces the options facing the population, and is likely to lead to a decline in the long

— *Birds:* Studies in Germany suggest that tree decline can result in a temporary increase in some endangered bird species, including the three-toed woodpecker (Picoides tridactylus c citril finch (Serinus citrinella), crossbill (Loxia curvirostrata), rock bunting (Emberiza da), black grouse (Lyrurus tetrix) and nightjar (Anthus campestris), by increasing the number of dead trees and the herb and shrub layer in managed forests. However, the research also suggests that a greater number of species suffer through forest decline (and in any case the problems of the species listed above were originally caused by forest management that eliminated several important stages in the forest succession).

Air pollution does not constitute a single problem, but presents an array of threats and opportunities to plants and animals.

There is no single "air pollution problem". A wide array of pollutants, acting at different times and in a wide variety of combinations, interact with natural and with other anthropogenic factors to alter ecosystems.

For example, acidification presents freshwater birds with a range of different threats and opportunities; some face immediate problems, some can adapt to a certain level of changes but not to extreme acidification, while a third group may even benefit.

In forest ecosystems, years of research effort have failed to find a single factor influencing trees, but rather a whole array of different stress factors, which may or may not play an important role in any particular decline:

— Research in the Cascade Mountains of the USA suggests that ozone depletion is resulting in a decline in amphibian populations through its role in increasing egg mortality. Experiments using filtered and unfiltered light on high altitude, shallow water pools found that egg mortality in the Cascade frog (Rana cascada) and the western toad (Bufo borealis) was 40 per cent, as compared with 10-20 per cent in the control, while egg mortality in the northwestern salamander (Ambystoma gracile) reached 90 per

— The predicted impact of global warming will be a net loss of biodiversity and ecosystem stability, particularly in some key habitats, such as boreal forests, mangrove ecosystems, cloud forests and some wetland and peatland habitats. Some of these factors may interact with acid deposition. For example, research in the Netherlands suggests that the predicted increase in prolonged droughts may cause additional damage to moorland pools because of atmospherically-derived sulphur compounds. Drying out in fens can cause fish deaths through acid surges, and invasions of plants such as the filamentous algae Tribonema minus and the rush Juncus bulbosus.

Pollution impacts are further complicated by the fact that in most situations pollutants are acting in the presence of other factors which themselves have an impact on ecosystems. Separating out the key, or most important, factors is often difficult. Contributory factors fall into three main types:

— anthropogenic factors: such as forest management systems

— natural factors: such as landform and soil type

— factors which appear to be natural but have been influenced to some extent by human activities: such as climate changes induced by global warming, and introduced plant diseases.

Air pollution effects are thus both more complex and more wide-ranging than simply assessment of the damage to a few individual species might suggest. Some species gain in a polluted environment, at the expense of what is usually a larger majority that decline. In the following section, some ecosystem responses to these changes are briefly outlined.

Ecosystem Responses

Responses to air pollution are not spread evenly throughout the world. The response depends in part on the nature, concentration and timing of air pollution, but also on the existing status and nature of a particular habitat. In the following section, some general points are made about susceptibility to air pollution, along with a brief overview of some environments that have proved to be particularly at risk. Some environments are particularly susceptible to air pollution damage.

Ecosystems are likely to be most at risk if they:

— are already on substrates with a low buffering capacity, ie a low ability to neutralise acids

— receive occasional, heavy doses of pollution

— contain key species that are vulnerable

In the following section, some of the key pollution-susceptible environments are identified and discussed.

Freshwater Ecosystems in Base-poor Areas

The water in base-poor lakes and pools receiving a heavy load of acidifying pollutants tends to become more acid, with a range of environmental effects. Analysis of the composition of populations of diatom algae found in lake sediments has allowed researchers to trace the course of acidification. Typically, acid deposition is neutralised by basic materials in the water until these are used up, then acidity rises sharply; the so-called "titration effect". In a few cases, acidification effects can occur episodically in relatively neutral or basic water, due to a sudden and temporary flush of acid. This can be caused by snow-melt in the spring, or by heavy rains following drought, which wash accumulated pollutants from trees and vegetation into water courses. These acid flushes can sometimes result in large fish kills.

Acidification has been identified from many areas of Europe, including southern Norway, Sweden, Finland, Denmark, Belgium, mid-Wales, Scotland It has also affected large areas of North America, including parts of Canada such as Nova Scotia, Ontarioand Quebec and some of the eastern

USA. Impacts on freshwater life increase with the level of acidity. Some species and groups disappear quickly with the onset of acidification, while others remain resistant to damage and may increase in the absence of competition.

Forest Ecosystems in Polluted Environments

Forests are affected by air pollution, although issues of cause and effect remain contentious. A decline in tree health due to air pollution was recognised at least a hundred years ago, and in the 1930s conifer plantations were abandoned in parts of northern England because high SO_2 levels inhibited growth. In the 1970s, a new form of tree decline was seen in the Black Forest in Germany, and similar changes have been found in much of Europe, including wide areas of Scandinavia. Monitoring suggests that about 25 per cent of European trees show serious signs of ill-health, and that the situation has grown worse in the last 20 years. Current surveys cover 34 European countries. In 1993, 22.6 per cent of the total sample suffered defoliation greater than 25 per cent, thus being classified as damaged. This included 20.4 per cent of all broadleaved trees and 23.9 per cent of conifers. Countries suffering particularly severe damage included the Czech Republic (53.0 per cent of trees suffering moderate to severe damage); and Poland (50.0 per cent). For 11 out of 12 common species, the proportion of damaged trees has increased significantly since surveys began in 1988. Amongst the worst affected of the common European trees were oak (Quercus), pine (Pinus) and spruce (Abies). In some cases, this has also led to the death of many trees in particular forests. Similar declines have been observed in parts including North America, Asia and the Pacific.

Decline symptoms include chlorosis (or yellowing) in leaves and needles; premature needle loss or leaf fall; deformation in leaf shape and size; changes in the tree canopy including thinning and development of "storks' nest" shapes in conifers; abnormal branching patterns, including downward tilting of secondary conifer branches, known as the "tinsel effect"; deformation in roots; disruption of natural regeneration; bark necrosis; and increased susceptibility to disease and pest attack. Most researchers now agree that air pollution plays an important role in this decline, along with a mixture of other factors including climate changes, management and pest and disease attack - the multiple stress hypothesis. Some of these factors are inter-related; for example in some cases air pollution can increase the chances of pest attack, or weaken trees so that they are more susceptible to disease. Air pollution may also be contributing to observed changes in climate.

Mountain Environments

Several research projects suggest that high altitude or montane environments will be amongst the first to show effects of acidification. Harsh climatic conditions limit plant growth, so that plants are unable to absorb additional atmospheric nitrogen, which then leaks into watercourses. Wind regimes tend to transport pollutants from surrounding areas up into mountains. Although pollution often decreases with altitude, pollution deposition can remain high because of greater precipitation levels. Research in Norway suggests that washout rates for pollutants such as magnesium, calcium and sulphates was 5-15 times higher in mountains than in urban environments. Ozone levels also increase with altitude according to measurements taken in the Bavarian Alps A series of measurements in North America and Europe suggest that tree growth rates have increased due at least in part to deposited pollutants.

Despite clear evidence for the fragility of upland ecosystems, these have received far less attention than either forests or freshwaters, probably due to the lack of strong commercial or sports interests in many mountain areas.

Whilst these two ecosystems have been the most carefully studied, evidence for important impacts also occur elsewhere, for example:

— *Peat ecosystems:* Most peat bogs and mires are already acid or neutral, and an additional acid load can cause serious changes to the ecosystem.

— *Heathlands*: Acidic or base-poor heathlands can undergo major changes as a result of air pollution. For example, excess nitrogen inputs to unmanaged heathland in the Netherlands has resulted in nitrophilous grass species replacing slower growing heath species.

— *Microhabitats on acid tree bark:* Lichens are likely to decline more rapidly on acid rather than alkali tree bark.

— *Plankton communities in the ocean*: Research suggests that increased nutrient loading, and eutrophication, is having an important impact on plankton in some areas. Part of this nutrient loading is thought to come from atmospheric pollution.

Air pollution tends to reduce biodiversity, but not necessarily biomass or primary productivity.

Studies have shown that biodiversity tends to decrease under conditions of acidification and heavy dry deposition of air pollutants. This tendency extends across all groups of plants and animals studied in sufficient detail to draw meaningful conclusions.

These losses usually represent a decline in rarer, more sensitive species. Their places are, on the whole, taken over by commoner and more robust species. In the case of plants, this can include many successful alien or weed species that are adapted to a wide range of soil and climate conditions, and which can evolve fast enough to offset the problems presented by pollution. Air pollution thus presents an additional factor contributing to the overall global decline in plant and animal biodiversity.The Swedish Environmental Protection Agency states that:

Acidification of both terrestrial and aquatic ecosystems almost always results in a decline in biodiversity.

Over the past few decades, habitat loss has been the greatest single threat to biodiversity and ecosystem stability in most parts of the world. As a result, conservation effort has been directed towards reduction of these threats through establishment of reserves and protected areas and also, more recently, through changes in management. However, establishment of conservation areas offer little protection against change from air pollution, and research has now shown that many "protected areas" are, in fact, being reduced in value through the impacts of air.

Indeed, protected areas may be particularly at risk. Recent analysis within Europe has suggested that conservation areas will suffer a disproportionately greater risk of pollution damage, as measured by critical loads, than the environment as a whole. National parks and other conservation areas have tended to be established on land that is less suitable for agriculture or other commercial uses, and thus often on acidic or base-poor soils, where effects of acidification are generally more acute.

Ecosystem managers are left with few options for addressing problems from air pollution. Pollution control is a matter for governments and industry, and frequently for international diplomacy. Those charged with management of individual sites, or even with a single country's forests or freshwaters, have attempted to take steps to reduce the problem at the site of damage.

Options remains fairly limited. Some foresters have attempted to plant more pollution-resistant trees, but success has been limited and this approach offers few advantages to other wildlife species. The other main option, and the one implemented most widely in Europe, is to attempt to artificially reverse acidification by the addition of basic materials to an ecosystem, usually in the form of solid lime.

Liming of freshwaters has now been practised in parts of Scandinavia for well over a decade, and more recently liming of forest soils has also been attempted. The ecological impacts of lake liming has been studied in a series of research projects in Scandinavia and in the UK. It has been shown that many species recolonise following treatment.

Fish species also often recover, and a study of 112 limed lakes in Sweden found that fish populations increased from an average of 3.2 species per lake before liming, to 4.0 after 2-4 years and 4.1 after 5-9 years. Other studies found little evidence of increased diversity, but the return of sensitive species such as salmon and trout. Post-liming growth increases have been measured particularly for perch, char and pike. Several invertebrate species and groups also return, including mayfly species. On the other hand, groups that increased under conditions of acidification, including Sphagnum and the goldeneye and mallard, have decreased. However, there are also a number of potential or actual problems with liming, in terms of ecosystem disruption, long term changes in water chemistry and the risk of causing further damage by the shock of a sudden change in pH.

The fundamental problem of liming, and of all other attempts to treat the site of pollution damage rather than the source of pollution, is that any gains are both partial and temporary, as long as pollution continues. Costs are borne by the receiver, rather than the emitter, of the pollution. Changes in financial circumstances, and in political will, can change the response.

There are already reportedly signs that the Swedish government, for example, is considering the reduction of its liming programme for economic reasons. To date, research suggests that air pollution has been involved in decline and extirpation of species, rather than in their extinction. This situation is likely to change in the future, for the following reasons:

— particularly susceptible groups in temperate regions will continue to decline until they become extinct, and for example this remains a real possibility for some lichens;

— pollution effects are already starting to be seen on a broader scale in some areas of the tropics, where both biodiversity and ecosystem fragility are greater.

Below, in Table 1, a summary is given of the wildlife effects outlined above.

Table 1. Summary of biodiversity impacts from air pollution

Group	*Summary of effects*
Algae	Blue-green algae are particularly susceptible to a range of air pollutants, and some species are at risk in polluted areas.
Lichens	Probably the single group showing the strongest responses to pollution, both from dry deposited sulphur dioxide and from wet acid deposition. Many local and some national extirpations.
Bryophytes	Similarly highly sensitive to many air pollutants, particularly in the case of tree-living or bog mosses.
Fungi	Many mycorrhizal fungi decline in acidified environments.
Pteriodyophytes	Evidence for decline in some fern and many club moss species in polluted air.
Flowering plants	Increasing body of evidence for decline, both through SO_2 pollution and on acidified soils.
Broadleaved trees	Many trees decline in polluted environments, due to the impacts of both air pollution and other stress factors: the multiple stress problem.
Conifer trees	As above.
Micro-organisms	Zooplankton decline in diversity in acidified waters, and soil micro-organisms decline in acid soils.
Soft-bodied invertebrates	Almost all lower invertebrates decline in acid waters. A decline in many species, such as earthworms, is also known in acid soils. Increasing evidence exists that molluscs and other species are directly or indirectly affected by air pollution on land.
Arthropods	Many crustaceans and insects decline in acid waters, although a minority of insect species thrive in the absence of competition. Fragmentary information suggests that many species are likely to decline on land due to air pollution although some, particularly aphids, thrive and increase in environments with a high SO_2 level.
Fish	Species show a range of responses to acidification, with some disappearing in slightly acid waters and others able to withstand even fairly severe acidification. Decline has occurred widely in Europe and North America.
Amphibians	Many species decline in acidified waters, primarily due to reproductive failure. A few resistant species thrive in the absence of competition.

Reptiles	No information has been found on reactions of reptiles to acidification or air pollution.
Birds	A minority of species decline due to food chain effects from losses, particularly in acidified waters, while other species have proved adaptable enough to cope with any changes. Others are apparently directly affected by SO_2.
Mammals	Despite much evidence for build-up of heavy metals and sulphur in mammals in polluted environments, the main effects noted for mammals come from food chain effects in species such as the otter and elk.

The only effective policy response to air pollution problems is to reduce pollution at source, through a cutback in demand, energy conservation measures, fuel-switching and technical pollution controls.

The biological impacts of pollution outlined above constitute a serious problem for biodiversity in many industrialised parts of the world. They are another, compelling, reason for reducing pollution at source. This needs to be achieved through an array of different strategies, including:

- reducing demand through energy conservation and social adaptations
- fuel-switching away from the most polluting sources
- technical pollution controls

At the moment, there are a number of worrying signs that governments are slipping in their commitments to previous agreements regarding pollution reduction, and are throwing up fresh obstacles in the way of further controls. Yet repeated public opinion polls show that people support pollution control. Firm guidance from existing internationall initiatives, including the Convention on Biological Diversity, could help focus governments' attention on addressing these issues at source.

References

Dudley, Nigel (1988); *Acid Rain and Wildlife*: A Report to Wildlife Link, Earth Resources Research, London.

Harter, P (1988); *Acid Deposition: Ecological Effects*, IEA Coal Research, London.

Tickle, Andrew, with Malcolm Fergusson and Graham Drucker; Acid Rain and Nature *Conservation in Europe: A preliminary study of areas at risk from acidification*, WWF International, Gland, Switzerland.

Ulrich, B and J Pankrath [editors] (1983) *Effects of Accumulation of Air Pollutants in Forest Ecosystems,* Reidel Publishers.

5

Climate Change and Biodiversity

It is now widely recognized that climate change and biodiversity are interconnected. Biodiversity is affected by climate change, with negative consequences for human well-being, but biodiversity, through the ecosystem services it supports, also makes an important contribution to both climate-change mitigation and adaptation. Consequently, conserving and sustainably managing biodiversity is critical to addressing climate change.

What is Climate Change?

In the atmosphere, gases such as water vapour, carbon dioxide, ozone, and methane act like the glass roof of a greenhouse by trapping heat and warming the planet. These gases are called greenhouse gases. The natural levels of these gases are being supplemented by emissions resulting from human activities, such as the burning of fossil fuels, farming activities and land-use changes. As a result, the Earth's surface and lower atmosphere are warming, and this rise in temperature is accompanied by many other changes.

Rising levels of greenhouse gases are already changing the climate. According to the Intergovernmental Panel on Climate Change (IPCC) Working Group I (WGI) Fourth Assessment Report, from 1850 to 2005, the average global temperature increased by about 0.76°C and global mean sea level rose by 12 to 22 cm during the last century. These changes are affecting the entire world, from low-lying islands in the tropics to the vast polar regions.

Climate change predictions are not encouraging; according to the IPCC WGI Fourth Assessment Report, a further increase in temperatures of 1.4°C to 5.8°C by 2100 is projected. Predicted impacts associated with such temperature increase include: a further rise in global mean sea level, changes in precipitation patterns, and more people at risk from dangerous "vector-borne diseases" such as malaria.

The present global biota has been affected by fluctuating Pleistocene (last 1.8 million years) concentrations of atmospheric carbon dioxide, temperature, precipitation, and has coped through evolutionary changes, and the adoption of natural adaptive strategies. Such climate changes, however, occurred over an extended period of time in a landscape that was not as fragmented as it is today and with little or no additional pressure from human activities. Habitat fragmentation has confined many species to relatively small areas within their previous ranges, resulting in reduced genetic variability. Warming beyond the ceiling of temperatures reached during the Pleistocene will stress ecosystems and their biodiversity far beyond the levels imposed by the global climatic change that occurred in the recent evolutionary past.

Current rates and magnitude of species extinction far exceed normal background rates. Human activities have already resulted in the loss of biodiversity and thus may have affected goods and services crucial for human well-being. The rate and magnitude of climate change induced by increased greenhouse gases emissions has and will continue to affect biodiversity either directly or in combination with other drivers of change.

There is ample evidence that climate change affects biodiversity. According to the Millennium Ecosystem Assessment, climate change is likely to become one of the most significant drivers of biodiversity loss by the end of the century. Climate change is already forcing biodiversity to adapt either through shifting habitat, changing life cycles, or the development of new physical traits.

Conserving natural terrestrial, freshwater and marine ecosystems and restoring degraded ecosystems (including their genetic and species diversity) is essential for the overall goals of both the Convention on Biological Diversity and the United Nations Framework Convention on Climate Change because ecosystems play a key role in the global carbon cycle and in adapting to climate change, while also providing a wide range of ecosystem services that are essential for human well-being and the achievement of the Millennium Development Goals.

Biodiversity can support efforts to reduce the negative effects of climate change. Conserved or restored habitats can remove carbon dioxide from the atmosphere, thus helping to address climate change by storing carbon (for example, reducing emissions from deforestation and forest degradation). Moreover, conserving in-tact ecosystems, such as mangroves, for example, can help reduce the disastrous impacts of climate change such as flooding and storm surges.

The climate is warming in most parts of the world. The global average temperature has increased by about 0.7°C and the European average by about 0.95°C in the last 100 years. It is estimated that temperature will increase by 1.4–5.8°C globally and 2.0–6.3°C in Europe by 2100. Precipitation patterns are more varied. In the last 100 years northern Europe has become 10–40% wetter and southern Europe up to 20% drier. These changes are projected to continue. The Intergovernmental Panel on Climate Change states that 'the risk of water shortage is projected to increase particularly in southern Europe.... Climate change is likely to widen water resource differences between northern and southern Europe.' Eight out of nine glaciated regions in Europe have shown significant recent retreat. From 1850 to 1980 glaciers in the European Alps lost about one third of their area.

In the past 100 years, European sea levels have risen by 0.1 to 0.2 m. Currently the sea level around European coasts is rising at a rate of 0.8 mm/year to 3.0 mm/year. In addition, extreme weather events such as droughts, heatwaves and floods have increased in Europe, while cold extremes have decreased. The extent and rate of current climate change exceeds all natural variation in the last 1000 years and possibly further back in history. The speed of change makes it hard for humans, other species and ecosystems to adapt.

The modern landscape also provides little flexibility for ecosystems to adjust to rapid environmental change. In contrast with historical migration responses, species today must move through a landscape that is increasingly impassable due to the widespread loss and fragmentation of habitats. Indeed, the IPCC states that in Europe, 'adaptation potential for natural systems is generally low' in part because 'Europe is predominantly a region of fragmented natural or semi-natural habitats in a highly urbanised, agricultural landscape.'Biodiversity is inextricably linked to climate. Changes in climate affect biodiversity and changes to natural ecosystems affect climate.

Direct Impacts of Climate Change

This section discusses the most important direct impacts that climate change could have on biodiversity.

Species Ranges

Climate change is likely to have a number of impacts on biodiversity—from ecosystem to species level. The most obvious impact is the effect that temperature and precipitation changes have on species' ranges and ecosystem boundaries. Any particular ecosystem consists of an assemblage of species, some of which will be near the edge of their ranges and others of which will not. Those at the edge of their ranges may need to move due to climate change. Species that are highly mobile or opportunistic are likely to benefit at the expense of those that are not. The invasion of alien species may be facilitated in some parts of Europe. Root et al. found that out of 1700 species reviewed globally, 81% of observed shifts in range were in the direction expected by climate change. In Europe, Bakkenes et al. predict that under existing climate change forecasts only 32% of plant species existing in an average 'grid cell' in 1990 would still be present in 2050.

In terrestrial Europe, trees and shrubs have replaced many Arctic and tundra communities. The treeline and the level at which alpine plants are found in Europe is moving towards higher altitudes. In Sweden, the treeline is predicted to rise by 233 to 677 m depending on the climate scenario and location. European tree species may be replaced by new species that are better adapted to higher temperatures and drought stress. Plants or trees that need low temperatures in winter to trigger bud bursting in spring may be adversely affected. In the Netherlands, thermophilic plant species have become 60% more common compared with 30 years ago, and coldtolerant species have declined.

European insects, mammals and diseases also demonstrate range shifts. For example, out of 35 butterfly species in Europe, 63% have shifted their ranges northwards by 35–240 km over the last century, whereas only 3% have shifted south. In Finland, macrolepidopteran species richness is expected to increase as southern species shift their ranges northwards. Conversely, the distribution of northern species, comprising 11% of Finnish species, may shrink. Thomas and Lennon report that the ranges of many British bird species have moved an average of 18.9 km northwards in the last 20 years.

In European seas, temperate species have migrated about 250 km northward per decade, whereas sub-Arctic and Arctic species have declined in number. More temperate species are now found in the North Sea. European seas have shown surface temperature increases, especially in isolated basins like the Baltic Sea and the North Sea. This has increased phytoplankton biomass, and moved the ranges of zooplankton species by up to 1000 km in the last few decades. Clark et al. predict that the North Sea cod population will increasingly decline due to higher sea temperatures. Shifts in ecosystem boundaries could mean that protected areas, such as the Swiss National Park, no longer contain the species and habitats they were established to protect.

The Pasterze Glacier has also retreated several hundred metres since the 1970s, thus affecting the Hohe Tauern National Park in Austria. Under existing static conservation paradigms, little emphasis is placed on changing patterns of biodiversity. And few protected area systems have been formulated with reference to climate change, even in countries where effects will probably be large. The World Wide Fund for Nature (WWF) argues that protected areas offer limited defence against problems posed by rapid environmental change, and that protected areas themselves will need to adapt to meet the challenges posed by global warming.

Flooding and sea level rise will affect species' ranges and ecosystem boundaries as well as threatening wetlands and coastal ecosystems. Man-made coastal defences across the UK are eroding faster than ever and destroying important inter-tidal habitats for birds. The IPCC states that in European 'coastal areas, the risk of flooding, erosion, and wetland loss will increase substantially… Southern Europe appears to be more vulnerable to these changes, although the North Sea coast already has high exposure to flooding.'

Changes in Phenology

Climate change is also causing shifts in the reproductive cycles and growing seasons of certain species. This can alter the frequency of pest and disease outbreaks. Research by Parmesan and Yohe on the timing of spring events, such as egg laying by birds or flowering by plants, showed that in 61 studies, the timing had shifted earlier by an average of 5.1 days per decade over the last half century. Likewise, Root et al. found that out of 1700 species reviewed, 87% of observed shifts in phenology were in the direction expected by climate change.

In Europe, phenological data shows an increase in the length of the growing season by about 10 days from 1962 to 1995. Leemans and van Vliet note phenological shifts in certain mammal, reptile, fish and bird species. For example, bird life cycle events such as migration and egg laying appear to be occurring earlier. Increasingly warm winters associated with the North Atlantic Oscillation influence the development and fecundity of red deer (*Cervus elaphus*) in Norway. And in the UK, amphibians are spawning nine to ten days earlier than they were 17 years ago.

Changes in Species Interactions

The impact that climate change will have on many of the more complex interactions (predation, competition, pollination and disease) that constitute functioning ecosystems remains largely unknown. However, some studies provide indications of what might be expected. For example, the indirect effects of climate change can be seen in a food chain in the North Sea where warmer water is driving cold water plankton further north. This reduces the survival of young cod and sandeels. A sandeel shortage may in turn have contributed to the large breeding failure seen in seabirds on the North Sea coast of Britain in 2004.

Changes in competitive ability may also emerge. For example, early leafing trees will get a two-week head start on their competitors in terms of growth, and thus occupy an increasing proportion of woodland. Numbers of long distance migrating birds in the Lake Constance region (a Ramsar site at the border of Austria, Germany and Switzerland) have declined, probably due to warmer winters increasing the competitive advantage of resident bird species.

Mismatches in timing between interdependent species may occur, especially when changes in some species are cued by day length, and others by temperature. For example, in one wood in Oxfordshire, UK, there is evidence that blue tit hatching no longer coincides with peak caterpillar numbers. Disease and predation may also increase. For example, numbers of native pests such as the green spruce aphid may increase due to milder winters, and non-native pests such as the pinewood nematode, gypsy moth and Asian longhorn beetle are now found in southern British woods.

Extinction Rates

Climate change may lead to a sharp increase in rates of extinction. Thomas et al. studied five regions of the world, and predicted that if the present rate of climate change continues, 24% of species in these regions will be on their way to extinction by 2050. This study indicated that for many species, climate change poses a greater threat to their survival than the destruction of their natural habitat. Evidence directly linking climate change and species extinction is difficult to procure, but at least one species: the golden toad of Costa Rica, may have become extinct due to climate change.

Effects of Extreme Climate Events

Climate change related extreme events such as disease, drought, fire or an El Niño are likely to increase, and can seriously affect biodiversity. These extreme events may affect organisms, populations and ecosystems more than gradual global or regional changes in averages. For example, Spain lost more than 485,622 hectares of forest to wildfires in 1994 and Italy lost 149,734 hectares in 1998. The 1976 drought in the UK severely affected tree health, and vulnerable species like beech took years to recover. About 15 million trees were blown down in the UK during the October 1987 storm. Broadmeadow predicts that higher mean wind speed and an increase in the occurrence of storms will make British woodlands more vulnerable to wind damage.

Variations between Regions, Species and Ecosystems

The impacts of climate change on biodiversity will vary between regions. The most rapid changes in climate are expected in the far north and south of the planet, and in mountainous regions. Unfortunately, these regions are also home to many species which have no alternative habitats to which they can migrate in order to survive, and species which cannot easily compete with new immigrating species.

In Europe, the greatest effects of climate change are projected for Arctic Regions, the moisture-limited ecosystems of eastern Europe, and the Mediterranean region. The IPCC states that 'the Arctic is likely to re-spond rapidly and more severely than any other area on earth, with consequent effects on sea ice, permafrost, and hydrology'. It adds that 'polar warming probably should increase biological production, but different species compositions are likely on land and in the sea, with a tendency for poleward shifts in major biomes and associated animals'. Theurillat and Guisan state that high mountain systems such as the Alps will also be particularly vulnerable.

Some species and ecosystems are also more vulnerable than others, particularly small populations or those restricted to small areas. For example, endangered birds in the UK, especially those living in fragile mountain habitats, such as the dotterel (*Charadrius morinellus*), ptarmigan and snow bunting, may have nowhere to move to and be lost from the UK forever. In mountainous regions of Europe, endemic tree species have been replaced by other species (such as spruce and pine), which have migrated upwards due to a number of factors including climate change. The IPCC states that in Europe, 'Tundra areas have practically no adaptive options available'.

Indirect impacts of climate change

It is not just climate change itself that can impact biodiversity. In some cases, the strategies that are adopted to mitigate or adapt to climate change can affect biodiversity. This section discusses the most important indirect impacts that climate change could have on biodiversity.

Classic top-down approaches to climate change often equate to large infrastructure construction projects. Projects designed to support adaptation to climate change are often associated with physical protection. For example, large sea walls may be built to protect against storm surges and floods. These are particularly important in Asia, and the Netherlands continues to invest millions in improving its sea defences against sea level rise and storms. Such projects often negatively impact biodiversity. Alternative adaptation options, such as strategic placement of artificial wetlands can benefit biodiversity. Other adaptation projects involve using pesticides and herbicides to control pests and diseases that might increase due to climate change. Use of such chemicals may damage existing plant and animal communities.

Projects designed to reduce global greenhouse gas emissions and thus mitigate climate change are often associated with large renewable energy schemes. These often have poor outcomes for biodiversity. For example, large hydropower schemes can cause loss of terrestrial and aquatic biodiversity and inhibit fish migration. Locating dams near river estuaries is particularly damaging, as it disrupts the whole watershed.

Dams can also be net emitters of greenhouse gases if submerged soils and vegetation decay and release carbon dioxide and methane. Europe accounts for 70% of the total installed wind energy capacity in the world. Wind farms pose three main problems for birds: disturbance, habitat loss/damage, and collision. For example, wind turbines in Tarifa and Navara in Spain have killed many raptors. The design, location and management of wind farms can limit these problems. The effect that offshore wind farms will have on sea mammals, fish and marine aquatic communities is largely unknown.

Afforestation and reforestation activities designed to sequester carbon and therefore mitigate climate change can restore watershed functions, establish biological corridors and provide considerable biodiversity benefits if a variety of different aged native tree species are planted. Monocultures, however, not only reduce biodiversity, but also increase the chances of pest attacks, thus challenging the permanence of carbon stocks. The location of afforestation and reforestation projects is also important. Replacing native grasslands, wetlands, shrublands or heathlands may lead to dramatic biodiversity losses, and also lower the relative increase in carbon sequestered compared to implementing such projects on degraded land.

Ecosystems and Biodiversity Affect Climate

Just as climate change affects biodiversity, so changes in biodiversity can affect the global climate. This section documents some of the most important ways in which biodiversity affects the climate.

Land use changes and subsequent changes to biodiversity can both increase and decrease greenhouse gas emissions. Countries like Ireland, the Netherlands and Denmark are significant carbon sources, but countries like Slovenia and Slovakia are significant carbon sinks. Forests are a major carbon store. Carbon dioxide is released whenever there are forest fires, or when forests are cut down. Globally, deforestation, mainly in tropical regions, is thought to be responsible for annual emissions of 1.1 to 1.7 billion tonnes of carbon per year, or approximately one fifth of human carbon dioxide emissions.

European forests were almost completely cleared during the agricultural expansion of the 16th to 18th centuries. This released considerable quantities of carbon from the vegetation and soil. European forests currently store on average 70–160 g of carbon per square metre per year, 70% of which is in trees, and 30% in soils. This makes them an important carbon sink. During the 1990s the European terrestrial biosphere stored 7–12% of the annual anthropogenic carbon dioxide emissions. Nabuurs et al. predict that this is likely to continue, as climate change will increase the quantity of stemwood in European forests with an additional 0.9 m3 per hectare per year in 2030.

Peatlands provide many environmental services, such as improving water quality. Many are important biodiversity reservoirs or stopover points for migratory species. Peatlands also hold roughly one-third of the soil carbon worldwide, and greenhouse gases are released every time they are burned, drained, converted to agriculture or degraded. Peatland forest fires in Indonesia during 1997 released an amount of carbon dioxide equivalent to 40% of the world's average yearly carbon emissions from fossil fuels. In Europe, peat extraction for horticulture, agriculture and energy is reducing stored carbon amounts. In Russia, permafrost peatlands appear to be melting and drying out and fire frequency is increasing.

Melting permafrost and the consequent release of greenhouse gases from northern wetlands can enhance climate change. Overall, European peatlands are a net source of carbon to the atmosphere of 70 Tg of carbon per year, equivalent to about 20% of the carbon sequestered by the European forest sector. Currently, some 60% of anthropogenic global greenhouse gas emissions originate from the generation and use of energy. While the use of renewable energy sources, such as wood, instead of fossil fuels can help mitigate climate change, this can also have negative impacts. For example, using relatively undisturbed natural fuel sources (such as native forests as opposed to plantations) can lead to significant biodiversity losses. Some bio-energy plantations replace sites with high biodiversity, introduce alien species and use damaging agrochemicals.

There are also many complex feedback mechanisms at work between biodiversity and climate change. For example, some species of ocean algae release dimethylsulfide into the atmosphere. Rising ocean temperatures due to global warming can lead to algal blooms, and resultant increases in dimethylsulfide release contribute to cloud formation. This in turn may help reduce temperatures, as less heat will be able to reach the Earth's surface.

Rising temperatures can stimulate the growth of phytoplankton in the sea, which can increase carbon dioxide uptake. Likewise higher atmospheric carbon dioxide concentrations and warmer temperatures can stimulate forest growth and hence carbon dioxide uptake in mid and northern European forests. In high latitudes, the replacement of shrub/tundra vegetation with trees can affect the radiation balance. Unlike snow, trees are more likely to absorb sunlight rather than reflect it. This in turn can enhance climate change.

Integrated Policy Responses

The many ways in which climate and biodiversity interact suggest a need for more integrated policy responses. Synergies between the United Nations Framework Convention on Climate Change (UNFCCC) and the Convention on Biological Diversity (CBD) need to be explored. The Ad Hoc Technical Expert Group on Biological Diversity and Climate Change has identified some possible linkages, and the convention secretariats have established a joint liaison group to help link initiatives relating to climate change and biodiversity. However, getting those responsible for implementing the two conventions to work together is difficult. The conventions have separate constituencies, administration arrangements, negotiators and guiding scientific bodies. Encouraging countries to set up a single institution to deal with obligations under all international environmental agreements could be one way forward.

The UNFCCC states that the ultimate objective of the convention is to ensure atmospheric greenhouse gas concentrations stabilise 'within a time-frame sufficient to allow ecosystems to adapt naturally to climate change'. Parties to the UNFCCC are guided by the principle that carbon sequestration activities should also contribute to biodiversity conservation and the sustainable use of natural resources. But common guidelines on what this timeframe is or how these principles might be realised have not been agreed.

The Kyoto Protocol and Marrakech Accords (under the UNFCCC) do not explicitly exclude practices such as afforestation of native grasslands or wetlands. The Clean Development Mechanism (CDM), established under the Kyoto Protocol, helps developed countries to meet emissions reduction targets, by allowing them to take credits from emissions reduction projects in poor nations. Projects are supposed to provide global benefits from carbon sequestration, but also sustainable development benefits.

These benefits could actively incorporate biodiversity conservation, soil protection and other environmental concerns. However, without a minimum set of common international standards, CDM and other climate change mitigation projects could flow to countries with minimal standards, thus adversely affecting biodiversity. The UNFCCC also currently classifies harvested forest products as emissions as soon as they leave the forest site. This fails to recognise the value of carbon sequestered in wood products and the biodiversity benefits of a well-managed forest.

Biodiversity-related agreements must also incorporate climate objectives. For example, many projects and policies to conserve and sustainably manage ecosystems undertaken by parties to the CBD and other biodiversity-related agreements (such as the Convention on Migratory Species, Convention on Wetlands and World Heritage Convention) could impact climate change mitigation and adaptation objectives.

Regional policymaking also needs improved integration of climate change and biodiversity concerns. Europe has various policies on each, and although suggestions have been made on how to integrate these concerns, policy coherence on these issues remains inadequate. For example, Dudley states that the Common Agricultural Policy needs reforming to support the protection of natural woodland in the UK in order that it can adapt to climate change. European forest policy has made some efforts at integrating climate and biodiversity concerns.

The Pan European Criteria, Indicators and Operational Level Guidelines for Sustainable Forest Management, adopted in 1998, aim to support carbon sequestration in conjunction with biodiversity conservation. For example they promote afforestation and reforestation with native species. The General Guidelines for the Sustainable Management of Forests in Europe under the Ministerial Conference on the Protection of Forests in Europe does likewise.

Hannah et al. point out that species range shifts will not respect political boundaries, so effective conservation management will require new regional collaboration. The Pan European Biological and Landscape Diversity Strategy, endorsed in 1995 by 54 European states, recognises the importance of ecological networks and calls for the development of a Pan European Ecological Network. The Habitat Directive of the European Union also acknowledges the importance of landscape elements that enhance connectivity, and other global and European policies such as the Bonn and Bern Conventions oblige parties to conserve and manage listed species and habitats. Whilst not overtly considering climate change issues, such policies help reduce the vulnerability of biodiversity to climate change.

Nationally, policies and activities that benefit biodiversity and climate change adaptation and mitigation need promoting and mainstreaming into various areas of national policy making. This will avoid a purely sectoral approach and help enhance coordination between different government agencies. But it is not easy as many climate change and biodiversity-related activities are located within the Ministries of Environment, which are traditionally relatively weak when compared to ministries dealing with finance or land use planning. At present, coordination among sectoral agencies to exploit potential synergies is poor. National plans to deal with climateinduced disasters could identify vulnerable ecosystems as well as vulnerable human settlements.

The development of a workable carbon intensity labelling system, pro-wood building and packaging standards and invigorated recycling programmes would help to maximise the climatic advantages of wood use. This in turn could provide biodiversity benefits if wood is sourced from well-managed forests. Afforestation and reforestation guidelines could be incorporated into national forest programmes and National Biodiversity Strategies and Action Plans. Protected area policies and other species conservation initiatives also need to become more aware of the effects of climate change and to include this in their prioritisation of hazards and risk management schemes.

Linking Climate Change and Biodiversity

The first step to address climate change must involve reducing greenhouse gas emissions. This requires lifestyle changes to reduce per capita energy consumption, and also technological innovations to increase energy efficiency and help the shift to more renewable technologies. The location of renewable energy initiatives, such as dams or wind farms needs careful consideration of biodiversity impacts. For example, the location of offshore wind farms in Germany is considering ecologically sensitive sites as well as wind energy qualification areas. However, climate change is already happening, and will continue to do so whatever steps we now take. We therefore need to look at how we can adapt our countryside to reduce biodiversity losses.

Many conservation practitioners in Europe recommend adopting a landscape approach to reduce the negative impacts of climate change on biodiversity. For example, the Woodland Trust in the UK advocates a move from focusing on protecting a few individual sites to a landscape approach. This would involve integrating protection, restoration and extension activities. Wide application of green corridors and buffer areas around particularly important woodland sites is needed.

Functionally diverse communities may be better able to adapt to climate change than functionally impoverished systems. Conserving genotypes and species along with reducing habitat loss, fragmentation and degradation may therefore promote the long-term persistence of ecosystems. In Britain and Ireland, the Modelling Natural Resource Responses to Climate Change (MONARCH) programme of research led by English Nature uses modelling approaches to help nature conservation policy and management practices adapt to climate change. Results suggest the need for a flexible dynamic approach that can adjust to the changing distribution of species and habitat types. In Europe, the ACCELERATES (Assessing Climate Change Effects on Land use and Ecosystems: from Regional Analysis to The European Scale) Project funded by the European Commission is also assessing the vulnerability of European agroecosystems to environmental change.

Larger protected areas, which cover a range of elevations, microclimates and ecosystems, will be less vulnerable, as species will be able to migrate to another safe habitat within the protected area if climate change adversely affects their present one. Connections between existing protected areas will be increasingly important, as will buffer zones around protected areas, landscape connectivity and management of areas between core protected areas.

Integrated watershed management can increase water retention and availability in times of drought, decrease the chance of flash floods and maintain vegetation as a carbon sink. It can also conserve watershed biodiversity. Avoiding degradation of peatlands and mires is particularly important for retaining soil carbon stocks and protecting biodiversity.

Certain forest management activities can simultaneously provide biodiversity and climate benefits. These include encouraging native species, increasing rotation age, low intensity harvesting, reduced impact logging, leaving woody debris, harvesting which emulates natural disturbance regimes, avoiding fragmentation, provision of buffer zones, corridors and natural fire regimes, and limiting the use of toxic chemicals and fertilisers. Biodiversity and climate benefits are also possible from agroforestry, revegetation, grassland management and agricultural practices. These include conservation tillage, recycling and use of organic materials, maintaining continuous ground cover, intercropping and reduced use of pesticides and herbicides.

Planting hedges reduces soil loss and landscape degradation from erosion (problems exacerbated by climate change-induced droughts, floods and winds) thus increasing soil carbon stocks and agricultural productivity, and benefiting biodiversity. The concept of becoming 'carbon neutral' is gaining popularity with many businesses, which wish to contribute to climate change mitigation activities by offsetting their carbon emissions. Likewise, many nations have committed to reducing their net greenhouse gas emissions under the Kyoto Protocol of the UNFCCC.

The CDM provides one mechanism for doing this. Projects designed to sequester carbon, and hence mitigate climate change, present opportunities to incorporate biodiversity considerations. For example, The Netherlands Forest Absorbing Carbon Emissions (FACE) Foundation has invested in forest restoration activities and indigenous tree planting in areas of Mount Elgon National Park, Uganda, that were previously degraded. The Ugandan Wildlife Authority has implemented these activities.

The aim is to offset Dutch greenhouse gas emissions from the foundation's clients, which include power generating companies and industrial and business clients in Europe. Certified Emissions Reduction credits would be awarded under the CDM. Tree Farms, a private Norwegian company, has also funded tree planting in the Bukaleba Forest Reserve, Uganda, in anticipation of the CDM becoming operational. Both these projects could have been improved if local community needs had been accounted for.

Using more wood products can sequester significant amounts of carbon, particularly if products have a long useful life and are recycled when no longer useful. If wood is sourced from well-managed forests, biodiversity benefits can also accrue. This is particularly true for European countries where over 90% of imports of roundwood and sawn wood are from other European states and most forests are managed sustainably.

For short life cycle packaging materials, substituting one tonne of virgin card for glass, plastic, steel or aluminium results in average savings of 1.1, 2.8, 2.9 and 4.1 tonnes of carbon dioxide respectively. Despite this, many people think using wood substitutes is better for the environment.

Several possible tools for integrating biodiversity and climate change concerns exist. The ecosystem approach could incorporate climate concerns, and Environmental Impact Assessments, and Strategic Environmental Assessments can be adapted to support broad uptake of environmental priorities.

For example, Finland is applying a Strategic Environmental Assessment approach in developing its national climate strategy. Using these approaches to assess climate adaptation and mitigation projects, however, might raise assessment and compliance costs and ultimately prevent beneficial projects from occurring.

Effects of Deglaciation on Marine Biodiversity

Currently planet Earth is warming. This warming is very unevenly distributed over the planet and many places are changing little or even cooling. The Polar Regions, in particular the Arctic and the Antarctic Peninsula, are experiencing intense and rapid warming. One of the main effects of the warming has been melting of the ice accumulated in these parts of the Polar Regions. The effects of climate change are many and diverse, and difficult to separate. That both extreme northern and southern regions currently have ice caps is unusual until recently in geological time.

A number of events led to the development of the polar icecaps. The ice masses covering the continent of Antarctica and large areas of the Weddell and Ross seas formed millions of years ago (mya) following a prolonged period of cooling. Following fragmentation of a former super-continent (more than 100 million years ago) other continents drifted away from Antarctica.

About 34 mya the Tasman Rise separated and the Drake Passage opened, enabling deep water currents to entirely surround Antarctica. The Antarctic Circumpolar Current (ACC) commenced and Antarctica became oceano-graphically and geographically isolated. This isolation and the falling atmospheric carbon dioxide (CO_2) levels led to the region's cooling, development and growth of the huge ice cap and ice shelves.

In recent years, to great scientific and public concern, there have been spectacular collapses of floating ice shelves, such as the Larsen B (eastern Antarctic Peninsula). Following global cooling the ice mass in the Arctic formed mainly to cover the ocean around the north pole. The northern polar icecap also covered Greenland and some of the smaller northern islands such as Ellesmere (in arctic Canada). Many ice masses throughout the Antarctic Peninsula and across the Arctic are retreating and are accelerating in their rate of retreat. It is important to remember that the Polar Regions have been deglaciating for thousands of years since the last glacial maximum.

Separation of what, in terms of ice loss, might be expected from the cyclical natural deglaciation and melting in response to rapid warming over the last two centuries is not straightforward. Patterns in seasonal fast ice are changing too. For example the extent, duration and timing of Arctic sea surface freezing in winter is changing rapidly. The consequences of this melting are likely to be increased freshening of surface waters, stratification of the water column, near-shore sedimentation and turbidity, more iceberg scouring, exposing new habitat, and in the case of land-based ice sheets (such as on Greenland) rising sea level. All of these can drastically influence the biota (living organisms) in the water column, on the seabed and in the lakes and on land in the Polar Regions. Potential effects on biota are further complicated by other changes in the environment such as warming itself, surface water acidification (because of raised atmospheric CO_2) and increased human activity such as bringing in pest species, pollution and fishing.

Investigation of potential impacts on biodiversity requires background knowledge of the recent richness and actual distributions of organisms in the Polar Regions. A swimmer snorkelling in tropical waters at a Pacific and an Indian Ocean shore would see quite a number of species in common.

Doing the same at a site in each of the Polar Regions would be quite different; there are no species in common at all. Famously polar bears live in the Arctic and penguins in the Antarctic, but on the seabed things are even more different. There are a number of reasons why northern and southern polar organisms have so little linkage with each other. This lack of connectivity is important to the nature of polar biodiversity. First, the two Polar Regions are completely separated by thousands of km of warm water and Antarctica has been isolated for so long that most Antarctic species only occur there (they are endemic).

Figure 1. Biodiversity of the Antarctic continental shelf. The benthic animals are a pycnogonan or sea spider (A), polychaete (B), asteroid or sea star (C), echinoid or sea urchin (D), ophiuroid or brittle star (E), holothuroid or sea cucumber (F), demosponge (G), ascidian or sea squirt (H), actinian or sea anemone (I), anthozoan or soft coral (J), hydroid (K), nudibranch or sea slug (L), gastropod or sea snail (M), bivalve or clam (N), brachiopod or lampshell (O), amphipod crustacean (P), isopod crustacean (Q), bryozoan (R) and three types of worm: nemertean (S), priapulan (T) and sipunculan (U).

Second, the two Polar Regions are also very different. The Antarctic is a large continent surrounded by ocean while the Arctic is sea nearly surrounded by landmasses. The Antarctic Ocean is much deeper, has no rivers flowing into it, is less polluted at its margins, and older. Third, species compositions of temperate and tropical environments have become increasingly more similar. In the last centuries humans have directly or indirectly introduced and established thousands of pest species, such that many particular types now live throughout the world. The Southern Ocean is the only marine environment with no known established invaders.

Fourth, the Polar Regions have very different levels of biodiversity. If the numbers of currently known species of each polar region are compared to global averages by length of coastline, continental shelf area or ocean area the Antarctic is rich in marine biodiversity (Figure 1) whilst the Arctic is relatively impoverished. This is, however, a very crude measure and hides some important details. The level of species richness varies considerably with the type of animal, for example the proportion of the world's sea spiders (Pycnogonida) and polychaete worms is very high in the Southern Ocean whereas barnacles, crabs, cartilaginous fish and marine reptiles are very poorly represented.

An alternative measure is that amazingly representatives of as many as 15 of 36 phyla (major animal types) can be seen in a single SCUBA dive to about 25 m depth – more than for example could probably be seen in any other environment anywhere-else in the world. Particularly common and abundant seabed animals, apart from sea spiders and polychaete worms are starfish, sea urchins, brittle stars and sea cucumbers (all echinoderms). Other mobile animals which are common are sea slugs, snails and clams (all molluscs), amphipods and isopods (all crustaceans) and nemertean worms.

Common sedentary animals include priapulan worms and sipunculan worms. Finally many sessile animals are also abundant such as sponges (poriferans), sea squirts (ascidians), sea anemones, soft corals and hydroids (all cnidarians), lamp shells (Brachiopoda), and bryozoans. In the water column the most famous inhabitants are the whales, seals, penguins and a major component of their food is the crustacean krill (*Euphausia superba*). Other common animals in the water column near the surface include jellyfish (cnidarians), comb jellies (ctenophorans), copepods, amphipods and mysid shrimps (all crustaceans), pteropods (molluscs) and arrow-worms (chaetognathans).

Actually, a swimmer or diver would see very little life in the very shallow waters (first 10 m) because of frequent catastrophic impacts of small icebergs. Also on most polar shores any intertidal fauna present is cryptic, and so they appear fairly barren of life. Biodiversity is dynamic in space and time, as are the icecaps, so to be able to consider the influence of one over the other, knowledge of more than just the current state of both is needed. For example despite the famous adage about penguins, the Arctic did have a northern equivalent, the Great Auk, but it was hunted to extinction by humans.

Historical Outlook

Strong warming and ice retreat signals are both apparent regionally, but the highest resolution record of historical temperature, spanning 760, 000 years in ice cores, shows that regional (and probably global) values are rarely stable—temperature is either increasing or decreasing. Over tens of millions of years Antarctica has changed from a Cretaceous greenhouse to a late Tertiary icehouse. Today the Earth is warm but it has been warmer, even in the late Pleistocene, for example in the last interglacial period just 120,000 years ago (120 kyr). It is the rate of change (warming and melting) which is unusual but even this has some historical precedence associated, for example, with rapid massive releases of methane clathrate, during the Eocene thermal maximum.

There is certainly some evidence that the West Antarctic ice sheet has waxed and waned considerably and may have even collapsed during the late Pleistocene. The ice cap covering the northern polar region is much more recent (from about 3.6 Ma/Pliocene, late Tertiary) than that in Antarctica, and also seems to be highly dynamic. A problem for interpreting effects of previous wide-scale deglaciation, which has clearly occurred in the Polar Regions before, is that the advancing ice leading to the last glacial maximum tends to obliterate most of the evidence of what was there beforehand.

In the build up to glacial maxima, the peak of an ice age, the ice covering the continent and continental shelf scours and scrapes the land and seabed. The scours from the last ice age can still be seen for example on the seabed around Antarctica on sonar images. As many ice ages have come and gone in the last few hundred kyr evidence of effects on biodiversity are often scarce. Subglacial erosion and its cyclicity all lead to poor fossilisation of organisms on the polar shelves. Around Antarctica just a few localities such as Seymour and James Ross islands have been found with a fragmented fossil record covering the last few tens of millions of years.

During the glacial maxima ice sheets would have extended far onto the shelf (as ice shelves) and bulldozed most life beyond the shelf break, maybe to water depths more than 800 m. This is crucial in context to biodiversity changes following the melting of ice caps, because it means that the continental shelf must have been recolonised from deep water each time when the ice sheets melt. This implies that only those species, which can survive down at those depths will be left to recolonise the continental shelf when the ice does melt.

Marine biologists recently found that some animals such as the sea-urchins have gradually lost species with certain characters like planktonic feeding larvae. They theorise that the species with such planktotrophic larvae find it hard to cope with living in deep water for long periods of time. However, taxa with planktotrophic larvae can recolonise shelves rapidly and quickly and achieve high levels of abundance when the ice caps melt. so there is a trade-off between strategies.

One of the most important points, which are under scientific debate, is how completely the ice sheets covered the continental shelf, i.e. whether there were pockets left uncovered by ice as refuges. Evidence is emerging that this was the case for at least a few lakes and land areas so it may be possible for the seabed, too. Skeletons of microscopic algae in an Antarctic lake show that some species, which withdrew during the ice age, have not returned yet. It seems that Antarctica's coastal biodiversity thrives on the continental shelf in warm interglacials (as the present time, but only a total of 10% of the last 760 kyr).

In glacial times it survives in refugial pockets uncovered by ice, or in deep water or in the southernmost parts of more northerly continental shelves such as around New Zealand or Eastern Patagonia. The high diversity of some shelf isopods supports this theory. Furthermore benthic communities were discovered in the Ross Sea at a distance of 22 km inside the shelf-ice edge. Another possibility, termed diachrony, is that ice advanced over the shelf in different parts of Antarctica at slightly different times. In such a scenario, all of the continental shelf could have been covered by ice, but at any one time there was a (different) part of the shelf, which was not covered, for organisms to survive on.

The biodiversity of the Polar Regions has not just moved up and down the continental shelf (or to a much lesser extent moved north and then back south), it has also changed with time. Biodiversity was significantly decreased in the end-Cretaceous mass extinction (~65 Ma) and there have been numerous smaller extinction events since then. After a rapid increase of species following the mass extinction there has been a gradual increase in biodiversity with just a few groups disappearing or almost disappearing from the poles, particularly Antarctica. Details of any patterns of biodiversity increase or decrease are difficult to find because of the poor polar fossil record and most evidence comes from just a few groups which preserve well, such as the molluscs, bryozoans, echinoderms and decapod crustaceans. For Antarctica additional evidence has come from looking at southernmost temperate fossils, such as in Patagonia and New Zealand.

In the long period of isolation Antarctica has developed a rich marine biodiversity which has, so far, proved dynamic but fairly robust to major disturbances from cyclic glaciations, major super-volcanic eruptions (such as the Toba eruption) and other more regional phenomena such as asteroid impacts. It has been isolated for long enough that most species are endemic but not long enough for many genera or families to be so. The Arctic, by contrast, is typically species-poor and most of its marine biodiversity has completely been geographically displaced and reinvaded with each big melt-back.

The Arctic is still in the process of recolonisation since the retreat of the last ice-age. Melting of ice caps is likely to influence biodiversity differently with geography and bathymetry so it is appropriate to discuss influences in a number of separate categories. For example, life on the seabed on the polar continental shelf, like on continental shelves in general, tends to be quite different in abundance, richness and composition to that in the deep sea. Antarctic continental shelf biodiversity is considerable and arguably the most important and pristine environment threatened by the climate changes expected. Its fauna mainly consists of endemic species; losses will also be losses of global biodiversity—and thus this seems the most important region to consider impacts on first.

Antarctic Continental Shelf Biodiversity

Most of the land, the shore and even the shallows around Antarctica appear as an icy desert, but from less than 100 m water depth down to hundreds of metres below the sea surface, life on the seabed can resemble a coral reef. Even in the shallows, areas sheltered from icebergs such as cliffs, overhangs and caves can be covered in animals. Parts of the Southern Ocean continental shelf have now been studied nearly as well as those at lower latitudes. However, the rate at which new species are being found and described in some groups (e.g. polychaetes worms or isopod crustaceans, sponges, bivalves and gastropods) is not even beginning to slow down. That is, the species description curve has not reached, or approached its asymptote (the point at which curves flatten out near their peak value).

Estimating the total biodiversity of animal groups with steep species description curves is almost pointless because it is largely guesswork. This strongly restricts any predictions of influences of ice-cap melting on these particular animal types, apart from perhaps at local (or maybe regional) level. There are also considerable areas which have hardly been sampled, which lead to similar problems. For instance, suggestions are hard to make on how the fauna of the Amundsen Sea might be influenced, given that what lives there is little known. The fauna that we do know, i.e. the thousands of species that have already been described from the Southern Ocean, comprises many unusual and impressive species.

It is not just that they are endemic that makes them striking. Some types are rarely seen away from deep-sea environments, many are very slow moving or do not move at all, some are very big (compared to the size of the same types elsewhere), conversely some are quite small and some are very old. Investigations of some of the towering sponges, clam shells and super-abundant sea-urchins have revealed almost no growth over long periods of time revealing them to be very old.

Scientists studying living lampshells found the isotopic signals of atomic bomb tests, done decades ago and thousands of miles away, incorporated into their shells. Some of these shells, little bigger than a person's thumb, may live to more than a hundred years old and store in their shell growth lines a history of environmental conditions every year for that time period.

Potential impacts on these old communities of old animals may take considerable periods for recovery. An example is how slowly populations of the great whales are taking to increase after being decimated by whaling. In many ways, trying to gauge how life on the seabed around Antarctica will respond to change is confounded by apparent conflicting evidence. For example, Antarctic communities and populations appear to be very old and fragile. Yet they live amongst catastrophic seafloor destruction by icebergs and survive for tens of thousands of years when ice sheets remove all (or nearly all) life on the continental shelves.

Experiments have shown Antarctic animals are highly stenothermal (can only tolerate narrow temperature changes). However, representatives of many groups have been found to live in the intertidal zone with extreme temperature changes, in warmer conditions at South Georgia and in Antarctica in the warmer past. A final group that should be considered are chemoautotrophic ecosystems under ice shelves. These communities are based around bacteria using chemical energy seeping from springs on the seabed, like those discovered around hydrothermal vents on mid ocean ridges.

Such chemoautotrophic ecosystems survive away from light energy, but their distinct communities are likely to be swamped by typical Antarctic shelf animals when overlying ice shelves collapse. We know almost nothing about these communities in Antarctica and it is currenty unknown whether any exist under ice shelves. Deglaciation and warming potentially threaten most Antarctic animals with change to their habitats beyond that they may cope with; effectively they may lose their habitat. In the short term though, the collapse of coastal ice shelves means new areas of continental shelf and slope become available for colonisation and thus expanding their habitats.

Changing Coastline

With the disintegration of some ice shelves and the retreat of coastal glaciers along the Antarctic Peninsula, especially at the northern end, the coastline is changing rapidly. The extent and rate at which this is happening has accelerated over the last half century. This is opening up new areas for colonisation and thus increasing the area of shallows for seaweeds and animals. Much of the coastline of the eastern Antarctic Peninsula (EAP) was, and still is, covered by an ice shelf. The ice shelf often extends into deep water (hundreds of meters) so a considerable stretch of coastline was nearly void of a rock intertidal or shallows. The collapse of parts of the Larsen ice shelf has massively increased the area of the intertidal zone, the shallow shelf environment and open water of the EAP. This is an important new habitat/environment, partly because of its size (which should increase as more ice shelves collapse further south) but also due to its position.

The EAP, unlike the WAP (Western Antarctic Peninsula), is on the edge of the Weddell Sea gyre and is bathed in cold, well oxygenated and dense water. The retreating ice caps will release a west Antarctic coastline of archipelagos. In Indonesia and Chile such coastlines have generated major radiations and abundance. There is some reason to believe that a fragmented complex west Antarctic coast could increase partitioning of the marine environment and so advance speciation.

If conditions do change to become too severe for some organisms in the northern WAP, they may retreat to cooler conditions along the EAP. In much the same way it is thought that many NW European species will or have started to shift their ranges northwards towards the Arctic, to stay within their climate niche. Away from the peninsular region nearly Antarctica's entire coastline is currently 'hidden' behind ice shelves, glacier tongues or ice cliffs.

Retreat or collapse of these would make a huge difference to colonisation opportunities for shallow and coastal biota. New data is showing that in places the water depth increases towards the continent underneath its coastal ice. There are many isolated deep (1 to 1.5 km) areas which will also be released when coastal areas deglaciate. At the moment, however, there is no evidence to suggest that the ice caps in East Antarctica are melting and their volume (most of East Antarctica is under more than 2 km of glacier) makes them robust to 'short' term variability in temperatures.

The sea and seabed around the edge of ice shelves can be highly productive areas. These centres of productivity are likely to not only become more dynamic but also to shift southwards as ice shelves collapse. What effect a southerly migration will have on these 'hot spots' is unknown. There is some evidence to suggest that it is not just ice sheets and glaciers that are melting back, the duration of winter fast ice (the sea surface which freezes seasonally) may also be decreasing. The duration and timing of fast ice is highly variable anyway but mean reduction of this is reduction of a habitat to organisms that live in its matrix, on or associated with its undersurface.

A major influence of fast ice to the global thermoregulation is surface albedo (changes in the degree to which light and heat are reflected). Reduction of fast ice cover per year in West Antarctica therefore, lets more light penetrate the water column and more heat exchange, so bringing changes to the water column. Across Antarctica though, winter ice cover is actually increasing by 0.5%. Compared to the Arctic, Antarctica's climate system is much more isolated from changes in lower latitudes.

Changing Water Column and Seabed

The water column in the Southern Ocean around Antarctica is typically deep, complex with many different water masses overlying each other, and seasonally very productive in some regions. The changing coastline, through retreating ice sheets and reduced durations of seasonal fast ice will have many influences on the water column, its biodiversity and the underlying seabed biodiversity. Clearly, there will be more and larger coastal areas open; potentially more light and heat can penetrate these. In addition, oxygen and carbon dioxide exchange will be enhanced, especially due to increased primary production (assimilation of CO_2 by phytoplankton = carbon sink). Evidence of the way the water column and organisms responded to warming and icecap changes in the last interglacial period is scarce.

As a result most predictions of change are conjecture, based on limited experimentation and there is a considerable diversity of ideas on how severely and quickly change will happen and which factors will be most important. From current patterns in the Arctic and emerging changes along parts of the Antarctic Peninsula other key factors influencing biodiversity are raising of sea level (in case of terrestrial ice sheets), freshening of surface layers, water stratification, increasing ice loading of coastal seas, and increased sedimentation and turbidity. Many of these have cascade ('knock-on') effects in which they influence many other things.

A good example of the varied cascade influences is that of ice loading. Icebergs come in many shapes and sizes, and are often carried by currents and wind and ground and gouge the sea floor, forming iceberg scours. When iceberg scours occur they create such heterogeneity to the seabed that scours thousands of years old can still be detected using multibeam sonar. Habitat heterogeneity itself can increase biodiversity, but the scouring causes massive destruction of the underlying community.

The injured, dying and dead provide important opportunities and food supplies for a diverse scavenging and predator community on the continental shelf. However, probably the most important influence on diversity is that ice scour opens up important new space (of all shapes and sizes) for colonisation. The colonisation of Antarctic seabeds follows a fairly predictable succession ending with 'dominants' that can monopolise space to the exclusion of other species.

The West Antarctic ice sheet has proved dynamic in the recent past, what is considered to be exceptional today is the rate of change. The massive ice sheet, which covers East Antarctica, holds much more fresh water than the Arctic, west Antarctic and elsewhere combined. If this starts to melt, which to date does not seem to have commenced, this would herald drastic changes to the Antarctic continent, the global ocean, global weather and undoubtedly to global biodiversity.

On one hand warming and even rapid warming has taken place before and is not believed by most scientists to have reduced the East Antarctic ice cap (although some glaciologists now think parts of it may have collapsed in the Pliocene). On the other hand the detailed record we have of atmospheric CO_2 and air temperature proxies over the last 760 kyr covering 8 glacial cycles shows that the link between these two factors is very good. Today CO_2 is higher than at any time in this record and rising rapidly.

Arctic Marine Biodiversity

The melting of the Arctic ice cap has been newsworthy and well documented because of its proximity to Europe and North America. Despite obvious inter-annual variability over the last 30 years in summer and winter ice cover in the Arctic, there has been a steady retreat of the summer ice cover each year. Permafrost is starting to recede rapidly in the high latitudes of Canada and Russia. Much of the media focus has been on influences on certain elements of the megafauna such as polar bears, seals and birds, which are losing their breeding and feeding habitats. For seals that breed on fast ice and animals that depend on snow and ice, the impacts of rapid habitat shrinkage are obvious and strongly negative, but these represent only a very small proportion of Arctic marine biodiversity. As in the Antarctic, human impact on many of the vertebrate megafauna in the Arctic has already been considerable (e.g. whaling, sealing and trapping).

Assessment of how marine biodiversity in general or even specific species are responding to Arctic deglaciation is not straightforward. For example a main predicted effect is southern species spreading up into the Arctic but ascribing this to recent polar ice cap retreat is strongly complicated by at least two factors. First, Arctic seas are considered to be still in the process of natural recolonisation in response to ice retreat since the last glaciation. Second, superimposed on slow natural invasion is much faster human mediated colonisation of non-indigenous (never naturally present) species (NIS).

On land, the melting of Arctic ice caps both facilitates new terrestrial species getting to the Arctic and establishing once there. Melting of floating ice is rapidly opening up the North-East Passage, which is likely to become 'permanent' in a few decades and this will certainly drastically alter species transport through and to the Arctic. Evidence from black shale rocks suggests that in the Cretaceous when no polar ice caps were around there was less circulation such that marine organisms may have had less transport opportunities. However, we consider that new organism colonists may be able to travel further and faster into the Arctic with the opening of a permanent seaway.

Certainly there will be more human transport such as ships carrying pest species clinging to their hulls or in their ballast water. Which of the new species would have got there anyway and which travel or establish because of climate change and melting will be very difficult to determine. The red king crab *Paralithodes camtschaticus*, deliberately introduced into the Barents Sea from the Pacific Ocean for fishery, spread across the Arctic into Norway, slowly at first but then very rapidly. The blue mussel *Mytilus edulis* has just been found on the coast of the Arctic island Spitsbergen (in the Norwegian administered archipelago of Svalbard).

The most recent fossils found to date of this species in the area show it was there before, ~8000 years ago. Species movements are increasingly more likely because of ice-cap melt back. How much influence ice-cap melting has had on species movements like these examples is unknown. Environments are constantly changing and the ranges of many species change in response. In parts of the Antarctic (e.g. Antarctic Peninsula or Victoria Land) species have the potential to migrate southwards in response to warming and ice melt, shifting their range to get back to their 'comfort zone'.

In the southern Arctic, species can move northwards along the coasts of Greenland or Canadian islands (such as Baffin and Ellesmere). However, most species present in the Arctic today are recent cool-temperate invaders (that have immigrated since the last ice-age from the North Atlantic or the North Pacific) on the northern edge of their range. Many of the current species which are successful in the Arctic are abundant because they are either pioneers (fast colonisers), fast growers and reproducers (to survive the frequent catastrophic iceberg impacts) or tolerant to cold, low salinities or near-shore sedimentation.

Most Arctic marine species live on the seabed, and for these species the effect of a shrinking polar region will change rapidly with time. On a short time scale of years/few decades, there will probably be reduced food supply and reduced life spans because of increased melt water, sediment and iceberg impacts (from ice shelf disintegration). Local and regional biodiversity will decrease. On an intermediate time scale, life expectancy of seabed animals will increase because of reduced iceberg scouring, and growth should increase because of increasing primary productivity reflecting a longer open water period and increased food supply.

At this time scale local and regional biodiversity is likely to increase, as more temperate species should be able to expand into the Arctic and many Arctic species should become more abundant. In little more than a century habitat for ice associated species may be critical and many seabed species may have reduced space to live because of increased competition and predation from new invaders.

The Arctic seabed may much more resemble that of temperate latitudes. On this time scale, local diversity is hard to predict, but will probably be increased regionally at the expense of reduced global diversity. This type of regional increase vs. global decrease has already started, associated with the 'global homogenisation' of fauna producing 'few winners, many losers' because many of the same pest species are invading and dominating regions around the world and pushing out or eating native species.

Cryptic encrusting communities (such as barnacles, bryozoans, polychaete worms and sponges) in the Arctic are an example of a group, in which biodiversity is likely to drastically decrease. These communities, which cover rocks and other substrates, are known to be organised in a highly hierarchical manner. They have a very strong 'pecking order' in which the top competitor smothers all others on meeting them in competition for space. The second best competitor is nearly always overgrown by the dominant species but overgrows all others. This pattern continues all the way down to the least ranked competitor, which is overgrown by all others but survives by reproducing fast, getting to unoccupied space quickly and growing rapidly.

The patchy nature of icebergs smashing into the seafloor constantly makes new space by destroying established communities and leaving bare rock. The seabed, thus, resembles a patchwork quilt of different species reflecting the different time periods since each area happened to be ice-scoured. In the future when large icebergs become rare, disturbance to the seabed will also become much reduced and as a result, a few dominant species may monopolise much of the space.

In contrast to the Antarctic, the Arctic basin has many large rivers flowing into it and thus seasonally has large volumes of freshwater 'lensing' on the surface. The rapid melt back of the northern polar ice caps has dramatically increased the level of freshening (lowering salinity) of the North Atlantic Ocean in the last few decades. Although there is high year to year and seasonal variability, oceanographers have found a distinct signal of consistent freshening of surface layers as well as increased acidification from absorption of atmospheric CO_2.

Arctic fauna does tend to be fairly euryhaline (tolerant of varying salinity) but such freshening strongly limits the establishment and survival of marine organisms in the shallows. Fresh water lensing of the upper water column also enhances stratification, which can alter water mass characteristics important to the underlying biodiversity. Of most notes, freshening and stratification can have a strong influence on the nature, diversity and duration of phytoplankton production. The amount and type of this primary production, which provides virtually all the food for both the water column and benthic animals, drives patterns of where, how and how many animals live.

Today's climate models are still a long way from supporting confident predictions about exactly how melting will proceed and in turn, how this will effect the environment. In part this is due to the strong greenhouse effect of water vapour, but past levels of cloudiness has not been calculated. Furthermore the simple addition of one of the complex network of passages around the Canadian Arctic islands can make a large difference to predicted ocean climate patterns. For example, adding the passage between Greenland and Baffin/Ellesmere Islands changed model outputs considerably. Despite the enormous complexity, at some locations around the Arctic we may be able to get insight today on the environments of tomorrow.

For example the east coast of Spitsbergen Is. is very cold and full of icebergs, whereas the west coast is much more ice free because it is warmed by the North Atlantic Current (extension of the Gulf Stream). Comparisons of west-east conditions and organisms may prove very valuable. If, as many oceanographers expect, the Gulf Stream and North Atlantic Current will weaken dramatically, western Spitsbergen (and indeed NW Europe) will develop considerably colder sea temperatures and more seasonal fast ice. In this case West Spitsbergen could come to resemble the environment of East Spitsbergen. Further warming and polar melting is likely to result in the biology of both coasts resembling the current east coast of the island and eventually the coast of NW Europe.

Deep Sea Organisms

Most of the marine environment can be described as 'deep sea', which covers 361 x 10^6 square metres, i.e. about 88% of the world ocean. Despite being rarely seen and poorly studied, the deep sea represents the largest habitat on Earth and makes a huge contribution to the heat balance of the World's climate system. There are only inadequate definitions about where the deep sea is; typically it is regarded as ranging from a few hundred metres depth (just deeper than the continental shelf) to the deep-ocean trenches down to 11,000 m. More than a century of deep-sea research has contributed a lot to the understanding of processes in the deep ocean, but with an average depth of 3800 m, it is clearly both time-consuming and costly to sample there. Hence, less than 1% of the deep-sea floor has been investigated to date. Even investigation on this small scale has revealed that the deep sea is not the constant ecosystem in time and space it was once thought to be.

On larger spatial scales, the deep sea is characterised by many topographical features, e.g. abyssal plains, seamounts, mid-ocean-ridges, hydrothermal vents and trenches (e.g. the Mariana Trench off the Philippines). Small-scale variations, such as physicochemical factors, food supply, water-mass transport, competition or substrate heterogeneity cause a patchy distribution of benthic organisms. However, all evidence to date suggests that the spatial variability in deep-sea areas is reduced compared to shallower waters and a crucial topic in deep-sea biology is to investigate the factors causing an amazing diversity of some taxa (e.g. peracarid crustaceans, polychaete worms, echinoderms) in the deep sea.

Apart from few exceptions, the physicochemical parameters, such as hydrostatic pressure, temperature, salinity and oxygen concentration are considered to be relatively constant. One strong causal factor underlying this constancy is that the water and currents overlying the deep seabed mainly originate in just two places, the Antarctic and the Arctic. Water which is very cold and saline, and thus very dense, sinks in the Weddell Sea to mix and form the Antarctic Bottom Water (AABW) and flows along the seabed of the Atlantic, Indian and Pacific Oceans, even north of the equator. At the other pole nutrient low the North Atlantic Deep Water (NADW) is formed in the Nordic and Labrador Seas, and spreads southwards. Where both these major water masses meet, the NADW (being slightly warmer and less dense) overlies the AABW. Oxygen can only enter the deep sea from the surface, so AABW and NADW essentially carry dissolved oxygen to the deep ocean.

The deep sea is an allochthonous system meaning its organisms are highly dependent on organic matter from the surface or water column above it, rather than from food produced *in situ*. This comes from sources as diverse as animal carcases, algae blooms, faecal pellets, and 'marine snow' (amorphous aggregates). Only in chemoautotrophic ecosystems such as hydrothermal vents or cold seeps is primary production generated *in situ* (autochthonous system) by bacterial symbioses.

Deep-sea organisms have a number of adaptations to help them sustain periods of limited and irregular food input. For instance, the scavenging amphipod *Eurythenes gryllus* is able to quickly consume large amounts (relative to its body size) when a food fall (e.g. a whale carcass) is found but also to withstand long periods without any food supply. Even on the deep seabed seasonality is evident, for example in reproduction patterns as seen in isopod crustaceans and echinoderms. The linkage between the photosynthetically productive water column near the surface and the ocean floor (often referred to as bentho-pelagic coupling) is strongly pronounced in the Polar Regions. This is because a high proportion of the primary production sinks straight to the bottom to be consumed by microbial life, mainly bacteria. Here, ice cover has an important role.

Formation of the sea ice increases surface albedo, so less light enters the water column, the rate of gas (crucially O_2 and CO_2) and heat exchange is limited and water mass circulation is restricted. Furthermore, the sea ice provides a self-contained habitat for a number of organisms living in and under the ice cover (e.g. diatoms, bacteria, amphipod and copepod crustaceans, krill), and facilitates an important flux of organic matter to the depth during the productive season. This cryo-benthic coupling, i.e. the linkage between sea ice and seafloor, is more pronounced on the shelf, especially in the Arctic, but also influences biomass and diversity of the deep-benthic fauna. Additionally, it is considered that released sea-ice nutrients may initiate an ice-edge plankton bloom influencing the grazing rates of, for example, krill, that provide food for higher trophic levels (e.g. squid, birds, seals, whales) and result in an increased flux of organic matter (e.g. faecal pellets) to the deep sea.

Deep-sea data is scarce and very little is known about ecological and evolutionary processes in this environment. As a result, little more than 'educated guesses' can be made about potential impacts of climate change to the deep sea. It is recognised that the current models are too simple, even to predict effects on the better known shelf seabed. Some scientists think the deep-sea environment of the Southern Ocean may hold a very high diversity of organisms, even comparable with coral reefs, whilst others assess it to have similar diversity to other deep-sea and (temperate) coastal areas. Estimates of the numbers of invertebrate species in the deep sea range from half to ten million and species description curves for many taxa (such as amphipod and isopod crustaceans) do not yet show any sign of approaching asymptote. It seems highly likely that such a major change to the system as melting ice caps will have a pronounced effect on the deep sea, which entirely surrounds the Antarctic continent, but we still may not understand the deep benthos well enough to be able to recognise organisms' response.

Possible effects of progressive polar ice melting include a more restricted circulation of the water column due to increased stratification. This would reduce vertical mixing between surface and deep waters and reduce exchange of oxygen and nutrients, while summer light levels would increase in the surface waters. Increased stratification may favour the growth of diatoms. Because these are heavily armoured with silica skeletons and often large, their productivity would be less usable to many benthic animals.

Some indication has led to suggestions that past periods of low oxygen and cooling events led to mass extinctions in the deep sea, e.g. in the Palaeocene and Eocene/Oligocene. It is assumed that some species survived this anoxia and underwent a period of intense evolutionary change on the abyssal plains. Skeletons of deep-water corals reveal that when large-scale ice melting processes occurred in past interglacial periods, the ventilation of deep-sea areas changed. Fossils of ostracod crustaceans and foraminiferans (minute small, shelled, single-celled organisms) from the North Atlantic show that the benthic species of these taxa were most diverse just before the glacial minimum, i.e. they increased coincidently with melting. It seems likely that some shelf species would avoid warming in the shallows by downward (submergent) migrations.

Recent investigations of some benthic groups (e.g. serolid isopod crustaceans) show that known Antarctic deep-sea representatives were formed by many separate, past and independent submergent migrations (at about 14 and 5 Mya). Such migrations may potentially be important in driving radiations (generating species) as well as extinctions. However, converse migrations, emergence from deeper areas to the shelf, also seem a possibility as niches become available. On the other hand, and in response to retreating ice caps, dropstones will be reduced normally providing oases for organisms needing hard substrates. This will cause a loss of niches and an assumed decrease of local and regional diversity.

Temperate and Tropical Shelf Biodiversity

Compared with influences on the Antarctic and Arctic, the melting of polar ice caps does not have such an obvious direct impact on tropical and temperate systems and their biodiversity. Many shallow tropical and polar invertebrates are believed to be similarly stenothermal—a conclusion supported by recent temperature tolerance experiments on bivalve molluscs, and so both are likely to be strongly influenced by rising temperatures. Melting of terrestrial ice sheets causes sea level rise (as does temperature directly through thermal expansion). This we know is strongly linked to mass extinctions from comparison with the fossil record.

Most of the data providing evidence of the connection between sea level change and mass extinction is from warm shallow seas. By contrast we have little information about how biodiversity has responded to sea level change in other habitats in the past and thus predictions for future response are weak. It seems likely that in the tropics, which are obviously most geographically removed from the Polar Regions, the principle influences will be rising sea levels and water-mass characteristics (such as flow rate and oxygenation).

Some coral atolls have failed to grow upward as fast as some past, rapid sea level rises and so the reef system has eventually died. In both cases, the short and medium term influences are likely to be regional decreases. Influences of ice cap melting on temperate biodiversity are probably more complex, particularly at higher latitudes. Furthermore the world's temperate seas are isolated from each other and in many ways quite different, again making generalised predictions difficult.

Some temperate coasts, like Patagonia and eastern Canada are being directly and strongly effected now because glaciers are retreating and changing the nature of the environment and the communities of organisms there. Other temperate areas, Southern South Africa for example, are less obviously directly influenced. Generally in temperate regions not only are rising sea levels important, but also, like in the Polar Regions, surface freshening, stratification and current and weather modification may have had big influences.

One of the most serious direct implications of ice retreat to temperate marine ecosystems is to the North Atlantic and North Pacific, and the effects will probably be considerable, maybe within just half a century. The 'temperate' ecosystems of North Atlantic and North Pacific are effectively expanding into the northern polar region. Change in geographic limits or their enlargement is not the biggest change that will happen, though. The melt back of the area of floating Arctic ice will create a permanent sea-way joining these two temperate areas.

When the Suez and Panama Canals were built, large-scale movement of non-indigenous species (NIS) took place. NIS (sometimes referred to as 'introduced' or 'exotic' species) are now widely recognised as one of the greatest threats to global biodiversity and the effects they have on native populations and natural environments can be seen nearly everywhere.

Climate Change and Biodiversity Conservation

The importance of protecting existing biodiversity cannot be overemphasised. Future biodiversity will adapt and evolve from the richness of the biodiversity conserved in the 20th century and from the extent of semi-natural habitats that will be protected, restored and created in the 21st century. But this is not to suggest biodiversity should be preserved in aspic. Climate change means that at sites currently considered to be of wildlife importance, some of the wildlife features, both species and habitats, may disappear, other species may arrive and habitats may change in their composition and structure.

Conserve Protected Areas and other High-quality Wildlife Habitats

In the 20th century an important focus for UK biodiversity and geological conservation has been the establishment of a series of Protected Areas to conserve protected wildlife habitats and associated species. Sites of Special Scientific Interest (SSSIs) and, in Northern Ireland, Areas of Special Scientific Interest (ASSIs), are the most extensive type of Protected Area for wildlife in the UK and cover about 10 per cent of the land surface.

A greater degree of control over conservation objectives is possible for these sites than for most others. The reasons for a site's designation generally include specified habitats and species. Consideration will need to be given to how designations can be applied more flexibly in future to protect changing and dynamic communities. High-quality wildlife habitatsmay occur outside of Protected Areas within areas managed primarily for purposes other than wildlife conservation, such as agriculture, forestry, water supply and recreation. In some lowland landscapes the extent of high-quality wildlife habitats outside Protected Areas is very limited: e.g. lowland heath. However, a large resource of high-quality wildlife habitats, notably ancient woodlands, moorlands and linear habitats such as hedgerows, streamsides and river banks, lies outside Protected Areas. Collectively these additional areas will play a critical role in the conservation of biodiversity as climate changes.

Two important functions of Protected Areas and other high-quality wildlife habitats will be as core areas for biodiversity and as connecting habitats within ecological networks. The early success of an ecological network will depend upon the persistence of large populations of as many species as possible in the core areas, capable of producing a supply of offspring to colonise new areas.

Conserve Range and Ecological Variability of Habitats and Species

Species and habitats are not likely to be at risk in all the places where they occur unless the risk is widespread.The risk of a species becoming extinct or a habitat being lost will be reduced if a varied set of locations are conserved which should encompass not only the full geographical range of each species and habitat but also the full range of ecological situations in which they occur.

For rare species and habitats it will often be appropriate to include all localities within conservation frameworks, such as action plans. For other species, bearing in mind that even some very common species may become vulnerable over time, sites with currently large populations are most important. Practically speaking, not all populations and habitat patches may be sustainable in the future and for those at the southern edge of their range, their maintenance should not be automatically set as long-term conservation targets requiring great effort but as a strategy for reducing the risk of extinction.

Species and habitats across the range of ecological situations in which they occur should be included, even where there are larger populations, or greater areas of habitat, in more typical conditions nearby. In the long term, these ecologically unusual populations and habitats may prove to be the places where the species or habitats persist.

Climate change is only one of many sources of stress to biodiversity in the UK.Like climate change, many of these stress factors have gradual but persistent impacts upon wildlife. Unless these threats are reduced or removed, action to combat the impacts of climate change is likely to be less successful. In addition, wildlife will be able to resist or adapt more successfully to climate change if these other stress factors are absent. While most are outside the immediate control of practitioners, it is crucial to be aware of their impact and take action where possible. Some examples of the more widespread threats are:

— Abandonment of traditional management, particularly in the lowlands;
— Over grazing, particularly in the uplands;
— Nutrient enrichment through use of artificial fertilisers, discharges of waste water, erosion and run-off of soil into freshwater and other sources;
— Introductions and spread of non-native species, some of which may change habitat structure, predate native species, are wildlife diseases, carry wildlife diseases or interbreed with native species;
— Intensification of farming practices, such as the creation of large fields that leave little space for wildlife, conversion of meadows to silage fields and increased herbicide and insecticide use;
— Over abstraction of water, resulting in low river or stream flows and alteration of water regimes in wetlands;
— Aerial pollutants, such as ozone, causing toxic effects and stress to vegetation especially in the uplands;
— Historical legacy of habitat loss and fragmentation, meaning that many species populations are now too small and isolated to be viable in the long term.

Ecologically Resilient and Varied Landscapes

Landscapes comprise a variety of habitats and an associated range of land uses. It is likely that landscapes will change in complex ways in response to both the direct impacts of climate change upon habitats and indirectly due to the way in which land use patterns alter or farming, forestry and water management practices adapt and change: for example, through increased irrigation or altered cropping patterns in agriculture. Within the landscape this may well result in some habitats increasing, decreasing or changing in structure, and others appearing for the first time or disappearing.

Maintaining a diversity of semi-natural habitats, increasing the area of semi-natural habitats through habitat creation and restoration, addressing the adverse impacts of unsympathetic land uses, and allowing natural processes to shape the ecology and structure of whole landscapes, will create the best possible chance for conserving the greatest amount of biodiversity. Reducing the intensity of land use in intervening parts of the landscape can greatly increase the chance for species to disperse between existing, restored and newly created high-quality wildlife habitats.

Conservation of Local Variation within Sites and Habitats

To wildlife the environment is a mosaic of potential habitat patches. Suitable patches are surrounded by varying amounts of habitat offering dispersal but not long-term survival possibilities, while other areas may be hostile. Each species requires a specific range of habitat-patch characteristics, including suitable climatic conditions, which allow it to establish, survive and reproduce. The size of habitat patch required to support a population of a given species varies enormously from a single leaf to many square kilometres. Climate can vary over very short distances – even two sides of a boulder can have very different conditions – and many species are highly sensitive to these very small differences, which are not readily perceived by humans.

Where there is a wide diversity of habitat patches species are more likely to be able to respond to climate change by relocating within the landscape they already occupy. In these cases only shortdistance dispersal may be necessary, so giving a higher probability that a population will be maintained than if long-distance dispersal is required to reach a suitable location. Those landscapes which are richest in terms of their current biodiversity are also more likely to be most varied in terms of their habitat mosaic and thereby most likely to allow some species to adapt to a changing climate by dispersing to nearby habitat patches in the future. The following characteristics are worth maintaining and enhancing:

a. *Diverse and structurally varied vegetation.* Different types of vegetation have different microclimates and some species may be able to adjust to climate change by expanding the range of vegetation types they occupy, or by moving from one type of vegetation to another.

b. *Semi-natural habitat on a range of slope and aspect.* Microclimate varies considerably with slope and aspect. At sites with varied topography species adversely affected by higher temperature and summer drought on south-facing slopes may be able to move to cooler, more humid northfacing slopes. Quite small differences in topography, for example on different sides of a hillock, may provide the topographical variation required if the magnitude of change is not too great.

c. *Uninterrupted semi-natural vegetation over a range of altitude.* For some species the response to climate change will be to move to higher areas, where climate is generally cooler and wetter. Hence, uninterrupted habitat within mountains and hills will allow the dispersal of species but montane species on the highest peaks are likely to be left with nowhere to go. This and the following point demonstrate the principle of connectivity.

d. *Uninterrupted semi-natural habitat across coastal zones.* Coastal areas have complex microclimates compared to inland areas and there is large climate variation over distances of less than one kilometre at the coast meaning species may find suitable nearby habitat patches as climate changes.

e. *Diverse water regimes.* Climate change is likely to have a complex effect upon water regimes. Summers are expected to become drier while winters are likely to become wetter and rainfall may become less evenly distributed, with more heavy rainfall events and flooding. The most complex range of habitats, and therefore the most aquatic and wetland species, are likely to survive in landscapes where there is variation from open water to dry land. A diversity of wetland conditions is most likely to persist where the open waters and wetlands are fed by a combination of surface drainage, ground water and aquifers.

Natural Development of Rivers and Coasts

The rivers and seas of the UK have an important influence upon biodiversity through the physical processes of erosion and deposition, which shape the landforms, and in turn the habitats, of the countryside and coast. In attempting to halt erosion and deposition and stabilise rivers and coasts by building sea walls, canalising rivers or otherwise artificially straightening and banking river courses, the already reduced biodiversity of coast and rivers has become more vulnerable to the impacts of sea-level rise and flooding.

On the coast, as sea level inevitably rises over most of the UK, formerly tidal areas will become permanently flooded. Where sea walls and other 'hard' features prevent the zone of tidal flooding from moving inland, the extent of intertidal habitat will decline and the associated wildlife will be lost unless suitable habitat is created or allowed to develop naturally through allowing tidal flooding further inland.

In the case of rivers an increased incidence of flooding is anticipated.Increased flooding has historically been managed by expensive dredging, straightening and embanking of rivers, a process which is highly damaging to biodiversity.

Making space for natural processes to take their course on rivers and coasts will be difficult to achieve in the short term as this will impact upon other land uses in the area, including built development in some cases. This approach may also mean that coastal freshwater habitats will be affected by saline incursions caused as rising sea levels cause change to their biodiversity. However, the alternative of attempting to delay the impacts of sea-level rise and increased flooding through engineering solutions is likely to become unreasonably expensive in many situations and technically impossible in others.

Creating Ecological Networks

Creating ecological networks that improve connectivity between habitat patches and allow species to disperse over larger areas will enhance the resilience of the landscape and further increase the probability of species surviving. In some landscapes a significant proportion of semi-natural habitats survive, most notably in the uplands and on the least modified coasts. Such relatively intact landscapes should continue to be an important focus for conservation activity because it is here that circumstances are most favourable for survival of biodiversity by dispersal.

The challenge lies in fragmented landscapes. Historic habitat loss means that in many parts of the UK, particularly in the lowlands, semi-natural habitats often survive as small, isolated patches. If species are to be able to move in response to climate change there is a need to assist dispersal through the wider countryside and colonisation of new sites. It is in these fragmented areas that ecological networks should be established and strengthened by programmes of habitat restoration and creation to improve opportunities for dispersal across landscapes and between regions in response to climate change.

Critical to the development of ecological networks is the conservation of Protected Areas and existing areas of high-quality wildlife habitat. These will form the core areas, rich in biodiversity, which will populate the rest of the network once connections are improved. Further complementary types of activity are required, firstly to restore habitat that has become degraded due to inappropriate management or abandonment on which there is a large body of practical advice availableand secondly to create new habitat, targeting it where there are greatest concentrations of existing semi-natural habitats.

Both restoration and creation are more resource intensive than conservation of existing high-quality wildlife habitat. Furthermore, in most landscapes relatively small areas contain the combinations of land use, land ownership and environmental characteristics that will permit habitat restoration and creation. But the benefits of this approach are now being recognised and this is an area where conservation practice is rapidly developing (e.g. the Wales Woodland Habitat Network Strategy and the Dutch Ecological Network).It is important to recognise that ecological networks can only enhance dispersal of some species and their development will reduce but not prevent biodiversity loss due to climate change.

Challenges of Biodiversity Conservation

Thoroughly analyse causes of change

One of the greatest challenges for nature conservation will be both identifying and responding to declines of species caused by climate change. It is important that all biodiversity change is not seen as an inevitable and unavoidable consequence of climate change. In many cases cessation of management, nutrient enrichment and other factors, often working in combination, will be the most important causes of habitat degradation and species decline. For some species, climate change may be only a small component of the problem but for others it may be the main cause.

Respond to changing conservation priorities

As resources for conservation are limited, conservationists have traditionally prioritised the conservation of rare and threatened species and habitats. This has been reflected by added legal protection and conservation focus for those species included in the UK Biodiversity Action Plan (UK BAP), Annexes of the EU Habitats or Birds Directives, Schedules of the Wildlife and Countryside Act or in Red Data Books. Site managers often focus their efforts on the basis of such status.

Under a changing climate, the range and abundance of many species will change, a process already well documented for many species groups. For some species and habitats special conservation measures may become inappropriate over time, as climate change means they become more abundant. A warming climate may offer many of the BAP species with southern distributions the potential to spread northwards and occupy a wider range of localities. Conversely, some currently common species may decline due to their sensitivity to climate change and will need to be prioritised.

A phenomenon almost certainly linked to climate change is the recent colonisation of the UK by species found on the near continent of Europe. Over time populations of these and other species newly found in the UK may decline in mainland Europe due to climate change, and UK populations may become of international importance. Hence, these species should begin to be considered in UK conservation target setting.

Integration of Adaptation and Mitigation Measures

An approach to conservation management on individual sites has been developed over several decades, which requires the identification of clear targets for habitats and species and the removal of obstacles to their achievement.With data from regular monitoring the effectiveness of management practices, such as grazing, can be reviewed, and if necessary amended. It is now essential that conservation management planning also incorporates consideration of climate change impacts at its initiation or review. There is a need to move from management largely focused on selected species and habitats towards much greater emphasis on the underlying physical processes that are essential to the maintenance of biodiversity on the site.

In the light of available evidence there should be an increased scrutiny in management planning of:

— Water regimes, as rivers, lakes, ponds and wetlands will require more active management in those parts of the UK where droughts are expected to increase in frequency;

— Fire control and management, as increased summer droughts may mean habitats in which summer fire is currently absent or rare could become prone to fire; generally summer fire may increase;

— Livestock management and cutting regimes, as the longer growing season may increase the amount of herbage available to be grazed, and earlier dates for cutting of hay meadows, verges and other habitats may become appropriate. In areas with increased summer drought lack of water for livestock and reduced plant growth may limit summer grazing or mean changes in the type of stock used;

— Increased control of alien species that threaten to become invasive as a result of climate change at an early stage in their establishment or spread.

In some cases these factors are under the control of an individual land manager, as for example grazing and cutting regimes. Many of the impacts that climate change may have on sites in the near future, such as summer drought and flood, are already encountered albeit less frequently, and suitable techniques for their management are available so that action can be taken.

There is likely to be more focus on mitigation through land use practices in future, particularly preventing loss of carbon and where possible increasing its storage. Most of the terrestrial carbon in the UK is in the soil and measures to reduce modification of soils – for example remedying the adverse effects of drainage on peat soils, or managing vegetation to minimise organic-rich soil disturbance – should be incorporated into management plans.

A large proportion of the carbon stored in UK soils and vegetation is within Protected Areas and other high-quality wildlife sites, especially in peatlands, and their management is a significant responsibility for land managers. There has, however, been very little quantification of the way in which conservation management practices impact upon carbon in the soil and vegetation. Such knowledge is urgently needed in order to ensure that they do not result in the increased release of greenhouse gases and, where possible, do result in carbon being removed from the atmosphere.

References

Hickling R., Roy D.B., Hill J.K., Fox R. & Thomas C.D., (2006)."The distributions of a wide range of taxonomic groups are expanding northwards", *Global Change Biology*.

Houghton J., (2002). *Global Warming the Complete Briefing*, Cambridge University Press,

Wallisdevries M.F. & Van Swaay C.A.M., (2006)."Global warming and excess nitrogen may induce butterfly decline by microclimatic cooling", *Global Change Biology*.

Clarke A. & Harris C.M., (2003)."Polar marine ecosystems: major threats and future change", *Environmental Conservation*.

Virtanen, T. and S. Neuvonen, (1999). "Climate change and macrolepidopteran biodiversity in Finland", *Chemosphere: Global Change Science*.

6

Impact of Tourism on Biodiversity

Tourism has become one of the largest economic activities in the world. The rapid growth in tourism has produced more infrastructure, increased pollution, put unsustainable demands on local environments, and created adverse impacts on biodiversity. In many coastal areas, for example, strings of high-rise hotels stretch for kilometres along beaches that not long ago might have been diverse habitats of coastal forest, mangroves, seagrass beds and coral reefs. The big challenges in minimising the impact of mass tourism on biodiversity lie with prob-lems such a sewage treatment in coastal areas, site selection of hotels, pollution caused by large fleets of tour buses, impacts of scuba divers, and environmental issues posed by golf courses, waste treatment of big resort hotels, etc.

According to the World Tourism Organization (WTO), world tourism in the year 2000, spurred on by a strong global economy and special events held to commemorate the new millennium, grew by an estimated 7.4 % - its highest annual growth rate in nearly a decade and almost double the increase of 1999. Nearly 50 million more international trips were made in 2000 - bringing the total number of international arrivals to a record 698 million.

The number of domestic tourists is still difficult to accurately quantify, but is estimated by some researchers to be as much as 10 times the number of international tourists. Clearly, tourism has a paramount economic role for countries around the world and, if planned and managed correctly, can significantly contribute to sustainable socio-economic development and environmental conservation.

However, inappropriate tourism developments—based mainly on the model of mainstream or mass tourism—are producing severe negative impacts on the natural and cultural environment, including biodiversity. If uncontrolled mass tourism is allowed to continue overrunning many areas of natural and cultural significance, irreversible damage will occur in these areas, which are the repositories of biological and cultural diversity in the planet as well as important sources of income and well-being for all countries and many local communities.

Even tourism developments in urban settings, far away from natural areas, may have unanticipated effects on surrounding lands and waters and the atmosphere, thus affecting biodiversity in many ways. Consequently, the appropriate interaction between biodiversity conservation planning and tourism planning and development has become a key concern for many institutions at the local, national and international levels.

The UNDP/UNEP/GEF Biodiversity Planning Support Programme (BPSP) has a mandate to provide assistance to national biodiversity conservation planners as they develop and implement their national biodiversity strategies and action plans. The integration of biodiversity into other sectors of the national economy and civil society has been identified as a critical indicator of successful implementation of sustainable development practices and of the objectives of the Convention on Biological Diversity (CBD).

To achieve this, UNEP has commissioned a series of thematic studies, each focused on one aspect of sectoral integration. One of these thematic studies is Integration of Biodiversity into the National Tourism Sector. Sustainable tourism has been highlighted recently as an area of major concern both within UNEP and the CBD.

The United Nations Environment Programme (UNEP) Division of Technology, Industry and Economics has, over the past two years, surveyed all main guidelines on sustainable tourism that are already available, and has consolidated and summarised these into a single set of proposed "Principles for Implementation of Sustainable Tourism". Outside of the mechanisms of UNEP and the CBD, a large number of other initiatives linking biodiversity and tourism (including ecotourism) have been undertaken by many organisations, ranging from the World Tourism Organisation (WTO), UNESCO, a number of NGOs, as well as numerous national and regional level destinations, and private tourism companies. It is extremely difficult for national biodiversity planners to (i) identify, (ii) access and (iii) assimilate all of this information on the interface between biodiversity and tourism.

It is also difficult for biodiversity planners to see the relevance of much "high-level policy-speak" to their day-to-day activities on the ground. Sustainable tourism has the capability of being a feasible tool for biodiversity conservation by providing economic alternatives for communities to engage in other than destructive livelihood activities, creating new revenue streams to support conservation through user fee systems and other mechanisms, and building constituencies that support conservation priorities by exposing tourists, communities, and governments to the value of protecting unique natural ecosystems.

Conservation Goals for Ecotourism

The International Ecotourism Society (TIES), coined one of the first and most well-recognized definitions for the term 'ecotourism' in 1991 which is, "responsible travel to natural areas that conserves the environment and sustains the well-being of local people." The process of introducing standards to this emerging profession is slow, because it involves multistakeholder involvement from around the world, including the involvement of peoples who live in some of the remotest parts of the planet.

The literature on conservation goals for ecotourism emerged in the early 1990s. The process of establishing principles for ecotourism in the 1990s primarily took place in conferences mostly in the U.S., Latin America and Australia often attended largely by academics and NGOs. This resulted in a variety of principles and standards that may or may not actually be applicable in practice.

The International Ecotourism Society found that guidelines offer a practical, participatory tool for coming to consensus on sustainable development practices directly with industry members. The guidelines development process was studied by TIES in the early 1990s to help determine how best to get a variety of stakeholders to adopt sustainability principles. TIES then adopted a standardized approach to developing ecotourism guidelines.

First, best practice information is gleaned from industry members via survey.

Second, participatory, multistakeholder meetings (that involved researchers specializing in impacts on natural areas and local people, industry, NGOs and communities) are held in northern and southern locations.

Finally, international review takes place, helping to assure that a wide variety of viewpoints from many nationalities are incorporated.

TIES has completed this procedure for Nature Tour Operators, International Ecolodges, and Marine Ecotourism. Certification of ecotourism is less advanced but also continues to develop for private sector products. The Ecotourism Association of Australia presently runs a national certification program with over 200 programs certified. This project is now being merged with Green Globe Asia Pacific and is testing ecotourism certification in a variety of destinations around the world.

Because ecotourism is a new category of assistance, few development agencies have specified policies to date. European researchers found that most development agencies link biodiversity conservation with poverty reduction within local communities, local cultural preservation, sensitive promotion to visitors and biodiversity improvements. However, there were few strict criteria and evaluation procedures. Only informal approaches for funding had been identified. This indicates the need for an immediate effort to establish a consistent evaluation framework that can be adapted for use by donor organizations around the world.

One of the key issues in development today, is how much the global marketplace can be transformed to be environmentally and socially sustainable. Ecotourism can be a form of commerce in which local peoples actively participate in the formulation of social "contracts" that go well beyond standard business formulas. Given that the large majority of ecotourism businesses are locally owned, and socially conscious owners are frequently involved, communities can achieve more than the status quo.

The involvement of the NGO and development assistance community also does frequently help to reverse top-down business patterns. Participatory decision making processes can make a difference, and increasingly these processes are being used, even in regions where communities have been sidelined in the past.

But at the same time, there are hundreds of rural communities, if not thousands, around the world that seek to be involved in commerce that will give them a fair chance to benefit, but do not have the capacity to do so, the access to training needed, or the chance to come to a reasonable understanding of what forms of commerce will or will not be beneficial to their community.

In the world of global and local development communities have been wrongfully exploited by outsiders seeking to use their expertise and knowledge without proper compensation in many instances not only in tourism but in all forms of enterprise.

Ecotourism principles, standards, and guidelines have evolved for private business but evaluation standards are only now being discussed for organizations carrying out ecotourism projects as a strategy for conservation. An informal inquiry sent to The Nature Conservancy, Conservation International, and RARE, resulted in finding that the development of evaluation and monitoring systems is still in the early stages.

The Biodiversity Support Network's final publication included ecotourism projects in a general framework of evaluation, which used indirect measures of conservation achievement. A "Threat Reduction Assessment (TRA)" approach was used to represent the conservation impact at each site. This technique examines the ability of the project to achieve biodiversity conservation by evaluating the area, intensity, urgency of each threat, as well as the degree to which all threats have been addressed by project activities.

An initial list of questions for ecotourism evaluation that would help donors and NGOs evaluate ecotourism as a conservation tool is given below.

1. To what extent has ecotourism contributed to the cost of protecting and managing natural areas/ or the specific natural area under review?

There is no international database regarding the use of visitor fees to parks, but anecdotal evidence indicates they have been introduced and/or increased at many developed and developing country natural areas in the 1990s. TIES assisted Colorado State University with a study published in 1994 that surveyed 319 protected areas in the world and found that over 50% of revenues for protected areas in developing countries was from visitor entrance fees. However to study real conservation impact it must also be noted exactly what portion of the fees are used directly in the protected area. In the 1994 study it showed that 32% of these fees were returned to the protected areas. Recent developments around the world indicate more tourism funds are being earmarked for conservation. The US Fee Demonstration Project provides 80% of the new fees worth a total of $176 million in 2000-2001 to the park or forest that collects it. In Belize, the Protected Area Conservation Trust (PACT) collects departure taxes for a fund used to finance natural area conservation. The proposed fee for PACT was reduced from US $10 to $3.75 because of opposition to fees from the tourism industry. Nonetheless, according the PACT 2000 Annual Report, US $500,000 per year in conservation fees are being collected per year, and between April 2000-March 2001 $275,000 in grants were awarded in amounts between $10,000 and $35,000 to both terrestrial and marine reserves.

It is widely felt that funding of public natural areas is not adequate around the world. In the early 1990s, IUCN estimated that 76% of protected areas budgets were not being covered by any source, and that an additional US $17 billion was needed to maintain the areas worldwide.

Evaluation questions should therefore include, how much total revenue is ecotourism providing to protected areas, what percentage is being earmarked for conservation in specific protected areas, and what proportion of total protected area budgets are being covered by ecotourism.

2. What are the biophysical impacts of tourism in natural areas?

The significance of biophysical visitor impacts on natural areas worldwide has never been quantified biologically. Tracking impacts is dependent on having or obtaining baseline data. But it is frequently unavailable. Giongo et al found in 1993 that monitoring of impacts was taking place in just over 50% of parks in developed countries and less than 35% in developing countries.

If biophyscial monitoring is not taking place, an evaluation framework can categorize the biophysical categories of impact and the percentage of area being affected. A recent study, using interviews with U.S. park superintendents in 51 parks, found 30% of the parks had significant impacts on vegetation, 37% on wildlife, 22% on water quality, 15% on air quality. This study did not inquire what percentage of the parks referenced were having these difficulties, a sorely needed piece of information.

The Giongo study shows that managers had concerns about erosion, site-spreading, trail depth, water quality and vegetation impacts. These problems were cited as being an issue in less than 20% of the parks in developed countries, and less than 5% in developing countries, again without information on the percentage of area being affected. Developed world parks had particular concerns about trail depth, site spreading and erosion, which were found to have exceeded acceptable levels in twice as many parks in developed countries than in developing ones.

Another evaluation focal point for the impacts of tourism in natural areas is the existence of direct management strategies, including zoning, required guides, citations and fines, campsite designation, limitation of visit duration, reservation systems, and visitor number limits. Ten years ago these strategies were used by less than 50% of developed world parks, and less than 40% of most developing world parks according to the 1993 study by Giongo.

The existence of regulations to manage visitation in parks and protected areas around the world may be an important indicator of progress in future to understand if ecotourism is having an impact on conservation of natural areas. However, Giongo notes that protected areas tend to avoid direct regulation of visitors as much as possible, preferring indirect management techniques.

Indirect management techniques include signs, patrols, tour operators, introductory talks, written material, displays, etc. These management approaches can also be tracked as part of an evaluation framework, to understand how well natural areas are using information to educate visitors.

3. Are there initiatives to manage ecotourism impacts via land-use management techniques and environmental regulations in buffer zones, or areas in the zone of influence of protected areas?

While some attention is generally paid to visitor impacts on areas within the borders of protected areas, startlingly little attention has been paid to the management of visitor impacts outside of the borders of protected areas.

The lack of controls on the use of land by both private landowners and commercial developers outside of protected areas can have a devastating effect, and this must be tracked in order to understand the impacts of ecotourism on natural areas. Very little data on this topic exists. Different approaches to managing tourism growth have been documented and these growth management strategies should be reviewed as part of a framework of evaluation when looking at regions affected by ecotourism development.

Zoning, or controlling quantity of visitors using a variety of zones designated according to type of use constitutes the most common approach to controlling visitor impacts, and can be considered to be a fundamental tool to protect areas from visitor impacts that should be tracked as part of an evaluation framework.

Carrying Capacity, or growth limitation strategies that set limits on numbers of visitors, parking spaces, hotel bed numbers, or ship berths, is another strategy that can be successful in managing the absolute numbers of visitors coming to fragile areas, and should be tracked as part of an evaluation framework.

Tourism plans are another strategy that can help the affected communities to understand what their goals are for protection, and allows community members to begin the process of establishing either zoning or limitation strategies.

Design planning can help areas establish the types of exterior elements desired, such as architecture, vegetation buffers, and density of each building site. This type of plannng helps the community to control its own sense of "self", and maintain control of outside development influences.

Regulatory mechanisms must also be tracked, such as the municipalities ability to control density of land-use, and what types of environmental impact analysis is required to receive building permits. In many areas municipalities still have no ability to limit commercial buildings that do not have sewage treatment or other basis health and safety requirements. Evaluation frameworks for ecotourism must include an evaluation of the municipality's ability to not only control growth, but to require proper sewage treatment and other fundamentals of environmental health and safety.

4. What impact is tourism having on biological diversity?

While tourism's impacts on wildlife and vegetation have been tracked by protected area managers in the past, there is no initial evidence in the literature of efforts to track tourism's impacts on biological diversity. Given that biological diversity is a primary source of concern at the global level, and that the conservation of biological diversity is often a goal of donor projects, some measurement of ecotourism's impacts on biodiversity should take place in order to measure the conservation impact of an ecotourism project.

Tracking impacts from the biological perspective may be extremely difficult. Baseline data would be difficult to obtain in most biologically rich areas. Adapting the approach of the Biodiversity Support Program , ecotourism projects could be analyzed according to their management structure, and then evaluated according to their conservation impact. For example, ecotourism projects could be categorized according to their management structure, such as community-based projects, private-sector/ NGO partnerships, private-sector/ NGO/community partnerships and so forth and then evaluated according to their conservation impact. As referenced above, the Threat Reduction Assessment Approach, constitutes the most likely framework for assessing conservation impact.

5. What impact is ecotourism having on the development of new government policies that support the sustainable development of tourism?

Evaluation of what government policies are in place for the sustainable development of ecotourism should be a part of any evaluation framework. Results from the TIES/UNEP Mesoamerica Preparatory Meeting for the Year of Ecotourism, a region that includes Mexico, Honduras, Nicaragua, El Salvador, Panama, Costa Rica and Belize, indicate that government policies are needed to achieve sustainable development and conservation results. The meeting conclusions include the following statement,

> National legislation is required to create the laws and regulations necessary to govern the development of ecotourism to guarantee the conservation of fragile areas and minimize undesirable impacts. A national legal framework is necessary to achieve appropriate regulations with clear legal guidelines, which can guide investment in sustainable tourism development and local infrastructure. National policies are needed to encourage rational use of renewable natural resources, to provide incentives for the use of renewable energy, and to encourage national and local investment into the appropriate management of solid waste and wastewater.

A possible framework of the evaluation of government policies could be derived from the box below written for a book published for the International Year of Ecotourism.

6. Is ecotourism contributing to a better understanding of the environmental and social setting of the site/region being visited?

The role of ecotourism in educating visitors and the community has been stressed throughout the world. Interpretation, which assists the visitor to gain a better awareness, appreciation, and understanding of and as a result, greater enjoyment from natural areas is an integral component of ecotourism. Understanding how ecotourism delivers information about the environment and local cultures is fundamental to evaluating it.

Tourism and Its Contribution to Sustainability

While defining conservation separately from sustainable development is an artificial division, in fact the discussion of how well ecotourism contributes to sustainability relies not just on a conservation bottom line, which has been argued by certain experts . Sustainable development literature has argued in the last five years, that the triple bottom line of conservation, economic, and cultural/social benefits need to be considered equally.

Ecotourism strives to be not only a conservation mechanism, and an economic development tool, but also a development process that seeks to remain harmonious with local cultural and social needs. Evaluating these concerns has been difficult indeed. But two new publications offer considerable assistance. Recent studies by the Organization of Labor in Bolivia, Peru, and Ecuador and the Pro-Poor Tourism Program are particularly outstanding examples of research that use frameworks for evaluation of case studies based on primary research that are clearly outlined and could be useful for other organizations as well.

Contribution to Local Business Opportunities

Tourism is a labor-intensive sector and has the potential of reducing poverty through employment. The Pro-Poor case studies evaluate complementary tourism enterprises as being equally important to the actual supply of the tourism products themselves, such as craft initiatives. Lessons learned from the Pro-poor tourism study show that credit and training are fundamental to the success of tourism as a tool to expand local business opportunities for the poor.

While business opportunities may expand, the pay for service can be extremely low, and that communities manage poorly the concepts of cost of goods, depreciation, amortization and the use of profits to build a business. Lessons learned from this study suggest that the training of community members in the management of funds could produce significant benefits and avoid possible conflicts among community members.

An evaluation framework could therefore include questions about credit availability for microbusinesses, and training for the management of small business finances. The value of labor could also be assessed in the area, and compared with what local community members are earning in the ecotourism project.

Bolivia offers very few legitimate examples of economic opportunity for local andindigenous communities in ecotourism, while Ecuador and Peru show increasing evidence of growth. The characteristics of products from these rural segments of society are often limited and quite dependent on outside forces, largely due to the centralization of the economies of the countries he studied. Nearly all capital is held largely in the cities with the economic elite. The marginalization and lack of attention to rural areas by state institutions in all three countries is a significant problem. As a result, the private sector is less likely to invest in rural areas because the potential for yields are reduced, by the lack of infrastructure. Nonetheless, the government of Ecuador has had a positive impact by investing in the legalization of community business, and assisting with community ecotourism development. The government Peru has assisted with community-based product development. Much of the NGO investment has been lost due to a lack of understanding of business development.

The pre-existing development scenario must be clearly outlined by a research process that seeks to identify the impacts of just one development tool such as ecotourism. In terms of inputs to this pre-existing situation, the involvement of government in developing business opportunities for marginalized populations is one significant evaluation factor. Other micro factors to consider would include the earnings of individuals that were previously below the poverty line, or may not have worked previously due to a lack of other economic activities in the area. Finally, it could be evaluated how many women gain opportunity through ecotourism and related businesses.

The collective benefits to be considered are numerous, and they go well beyond income and employment generated. Understanding collective benefits will help the evaluator to understand if ecotourism is contributing to the well being of local people.

The following categories should be considered for collective benefits:

— Human capital: skills, education, and health

— Physical capital: roads, water, and other infrastructure and tools

- Financial capital: credit and collective income
- Social capital and community organizations
- Access to information
- Policy context
- Market opportunities, livelihood options
- Cultural values
- Optimism, pride and participation
- Exposure to risk and exploitation

The communities confront many restrictions due to the structure of society, the market and the State, where they continue to be excluded and discriminated against in terms of their access to resources, public services, development opportunity, education, professional training, and health. All of this results in their lessened ability to genuinely take part in ecotourism or any development opportunity.

Social and Cultural Contributions

The evaluation of community based ecotourism via the application of sensible, rapid, and inexpensive evaluation and monitoring tools in the short-term that permit the measurement of economic, social, and environmental impacts is needed. At present few donors provide assistance or mechanisms for monitoring, and the open marketplace does not require the use of social or environmental impact monitoring tools.

The following factors should be considered while planning ecotorism strategies:

- Creation of standardized methodologies, adapted to local realities that include social, economic and environmental impacts.
- Provision of education and training for the delivery of tourism products that is equitable,
- Full respect and protection of the values, symbols, and cultural expressions of the communities' identity, language, customs, and traditions
- Strengthening of the organizational abilities of the community for representation at a regional and national level

— Development of a process of exchange between communities to enable them to develop a strong sense of solidarity with other communities and cultures around the world

— Strengthening, nurturing and consciously encouraging communities' ability to use traditional skills

The potential well-known negative social and cultural impacts of tourism must also be evaluated with feedback mechanisms for project redesign, such as:

— Loss of local traditions

— Commercialization of local products

— Erosion of self-worth

— Undermining of family structure

— Loss of interest in land stewardship

— Fighting among those that benefit and those that do not

— Crime and adoption of illegal underground economies

While the rights of communities to reject development is paramount, a great many communities seek access to information and have a desire to be less marginalized via participatory community, municipal, and civil society processes that include them. Ecotourism could be evaluated according to the kinds of "strategic alliances" that the community has gained as a result of its involvement in tourism, either with the private sector, NGOs, with other communities, or civil society associations. Ecotourism could also be evaluated according to the communications mechanisms brought to the community such as short wave radios or computers or Internet or other tools that give the community access to technical information desired or needed for development. There might be a review of how many community members actually take part in community, municipal and state level meetings in order to determine if ecotourism has enhanced or decreased the participatory nature of the local culture and society.

Biodiversity and Sutainable Tourism: A Case Study

Let us describe a case study of Fiji for linking tourism with the environment, biodiversity and sustainable development.

Fiji is a small country in the South Pacific, with over 300 islands and nearly 1 million people. Over half its population are indigenous Fijians.

Approximately 40% of Fiji Islanders are Indo-Fijians, whilst the balance is a mix of Chinese, European and others. Whilst Fiji has a strong sugar industry, and agricultural base, this is declining in importance. Tourism earned F$550 million for Fiji last year, and the Government is moving towards $1 billion as its target in 5 years time. In order to achieve the necessary growth, in both revenue and tourism numbers, to hit $1 billion there are significant hurdles to overcome. Not the least of these is the investment in infrastructure by Government, and the implementation of a development framework that ensures the protection of Fiji's natural assets.

Tourists come to Fiji because it has an extraordinarily rich biodiversity, it has a living culture, it has beautiful pristine marine resources and great rainforests, and it has the friendliest people in the world. Notwithstanding some recent political disharmony, it now has a stable, democratically elected government, which is focused on rebuilding the nation and on reconciling the ethnic divide.

As tourism operators in a rural area of Fiji, Turtle Island needed to establish its legitimacy in its community. Richard Evanson bought Turtle Island and moved there in 1972, and on the day he arrived, he started planting trees on his island, and creating relationships with the people in the 7 villages on the 4 other islands in the immediate vicinity. When he chose to open Turtle Island as a 5-star resort in 1980, he publicly acknowledged that one of the enduring goals for Turtle was to be a vital resource to its community. Richard's view has always been that as a tourism operator, if he did not add value to the lives of the people who lived in the community around him, then he simply had no right to be there. He wanted to strengthen the community in every respect, without impugning the integrity of their traditional and cultural bonds.

Over the years, Richard worked up his criteria of what would make a strong community in his area of the Yasawas. He believed the community needed to be able to:

- exercise independence in its decision making, planning and building process;
- be confident about its future, its sustainability and the prospects for its children;
- have leadership that leads from the top down, with Elders and other sectional leaders living and embracing the community values;

— have strong traditional and cultural bonds which are consistently reinforced to become part of everyday life;

— provide relevant, accessible and empowering education for future community needs; and

— be confident in a healthcare system that provides at least a "safety net" coverage to all.

From early on in the operation of Turtle Island, Richard set about the planning as to how he could assist the 3,500 people in the Tikina to achieve these goals to build a strong community. In doing so, he committed his own resources and energy, as well as harnessing the opportunities that are presented to Turtle Island weekly from its well heeled, socially concerned guests.

Although Turtle Island didn't realise it at the time at which the program commenced, Richard has been a visionary in the implementation in the field of Travellers Philanthropy enabling guests to engage with the local community, and connect with the place in such a way that some of them become inspired by the opportunity that exists for them to make a difference. He has created the vehicle whereby Turtle Island can act as a conduit for those guests who wish to contribute money, time, opportunity or goods to enable real benefits to be derived from that relationship.

The community benefits, and the guests benefit from the creation of that specialness between themselves and the people of the Yasawas. One of the central features of this has been the Yasawas Community Foundation (formerly known as the Turtle Island Community Foundation) to which guests can contribute, and (if US citizens) obtain a tax deduction. The Trustees of the Foundation include local chiefs and successful national figures from the local area (the Catholic Archbishop of Fiji is the Chair). Richard has also created opportunities outside the Foundation for guests to fund specific needs of the community if they arise.

In a more holistic sense, Richard has sought to use a three-pronged process in order to undertake his community capacity building. This includes implementation of programs in education, healthcare and job creation.

Education

Without literacy, numeracy and accessible opportunities, the community simply cannot prosper. In a competitive global world, Fiji has a lot of educational ground to make up, and there is an even greater need in the Yasawas. There are 3 primary schools operating in the area but no secondary schools. In order to move onto a secondary school after age 12, parents needed to send their young people off to town, to be boarders at one of the mainland secondary schools. With annual incomes in the region of approximately F$1,500 per family, and a cost for boarding school of $500 per annum, school retention rates after primary school are very low. In some of the primary schools, less than 50% of the 12 year olds go onto secondary school. The social impact of this lack of education cannot be ignored. Young people who don't go to secondary school are sitting around the village with not too much to do, with the potential of drifting off to the urban areas to find employment, or trouble. In 2002, Turtle Island founded its own secondary school. It started with 7 young people, in a makeshift classroom, employing a Fijian teacher.

Various supporter companies and guests in Fiji and the US were approached, and 7 computers were obtained, which were then networked, so that each student in the class had training from their first day of secondary school in IT. In 2003, a further 15 students are joining the school, with the original 7 students moving up to the next grade. There will be two teachers this year, and a better school facility is being constructed. By undertaking this project, Turtle Island has enable a few local students to remain in the area with their parents, to give them continuing schooling which many would not have been able to access, and have also provided them with consistent computer training that would simply not be available in a Fijian school. All of this is done at no cost to the parents.

By 2005, Turtle Island will have built its own fully operational secondary school, offering 5 years of secondary education and involving over 100 local children. In accordance with "religious practices" in Fiji, there will also be a rugby pitch out the front of the school!! This project is being financed by Turtle Island, but with assistance from some of its guests who are enthusiastic about the program, and inspired by its possibilities for delivering social equity.

Healthcare

In Turtle Island's area of the Yasawas, the medical facilities are, at best, rudimentary. No doctor has been posted to the area for over 7 years, and, whilst there are 3 nursing stations, they do not have power, running water or any significant expertise. Residents in the 7 villages, including young children, are dying or suffering badly from perfectly treatable diseases such as diarrhoea, within one mile of the Resort because there is not the expertise. Local people are simply used to going without treatment. It was evident to Richard that there was a lot of work that needed to be done to provide basic healthcare. There was not even a "safety net" in place. The first thing Richard Evanson did was to institute a series of medical clinics, and last month the Resort shut down for a week in order to conduct its 13 th Annual Medical Clinic.

The initial clinic that was established in 1991 was an ophthalmology clinic. Cataracts are a major problem in the area, leaving many islanders prematurely blind. From 1991, a specialist eye team has been coming to Turtle Island and undertaking cataract and pterigium operations. Turtle provides them with bures, and for the last 8 years, closes the resort whilst the Clinic takes place. This year, 18 members of the medical team were involved, with over 800 pairs of spectacles being given away to screened patients. In addition over 100 cataract and pterigium operations were completed, as well as 5 corneal transplants. Of these patients more than half were blind, and thus were able to get their life back again.

Since 1995, other specialist clinics have been introduced, for the local people, each of which operates for a week annually. These include dentistry, paediatrics, women's health and dermatology. It is essential that the community of 3,500 people has a fulltime doctor, and some reasonable medical facilities in which the doctor can work. Some of this money will be raised from US philanthropics. Much of it however, will be raised from the guests of Turtle who passionately believe in the need for proper healthcare to be provided to their friendly hosts, whom they have come to love whilst guests at Turtle Island. Turtle Island are also partnering with the Fiji School of Medicine, a number of specialist doctors from throughout the world, and the Fiji Ministry of Health to facilitate the construction of this Medical Centre.

Employment

An essential part of community capacity building is for young people to believe that there is a purpose to their education, their retention of good health, and that there is a career opportunity for them at the end of their educational road. Unfortunately, in so many rural and regional areas of Fiji, that career path is not obvious. In the Yasawas, tourism represents the only real option for employment.

One of the great natural assets of the Yasawas is its truly stunning white beaches, beautiful unspoiled villages, its crystal clear water, and its rich biodiversity. Three years ago, the High Chief of the area, whose people make up the population of the 7 villages, approached Richard, to ask him to help them build a budget resort on their island, on one of the most beautiful beaches in Fiji. He was concerned about providing jobs for his people, and saw tourism as the way to do it. Richard agreed to build and fund the resort, as a "social entrepreneur", on the basis that the budget resort was owned by the High Chief and the village, or mataqali, that he leads. It was agreed that the capital investment would be paid back from the profits on an interest free basis. It was also recognised by the High Chief however, that if the villagers tried to operate this on their own, they would not succeed.

Two years later, the project, known as Oarsmans Bay Lodge, is fully operational. The resort accommodates 56 people in a number of bures and dormitories situated along this beautiful white sand beach with clear blue water. The resort enjoys over 90% occupancy and employs 42 villagers from the High Chief's village, where previously there were no jobs. There are also other correlative benefits in the purchase of fish, vegetables etc from the villages. It's currently making a profit of between F$5,000 and 10,000 per month, which is gradually reducing the initial loan of F$700,000. In a few years, the village will completely own the resort.

Monthly board meetings are held with agendas, minutes, accounts, and the General Manager (a Fijian) is required to put forward his management reports and be accountable for them. The clientele is principally European budget and backpacker travellers, who read about Oarsmans on the Internet. The success of Oarsmans Bay has lead Turtle to be involved in two other ventures with other villages as social entrepreneurs. One of these is already operational, and the other will hopefully be operational later this year. These are based on similar partnership arrangements, with the end result being to provide jobs where they currently do not exist, and to have the resorts owned by the local people. The ultimate outcome Richard seeks is to create 500 new jobs, both directly, and indirectly, in the next 3 years.

Tourism Association

Prior to the opening of Oarsmans Bay Lodge, there were 7 or 8 small backpacker facilities operating in or nearby local villages, but they were not experiencing great success and certainly they had no commitment to sustainability or environmental best practice. Richard identified the need to bring together all of the operators who were seeking to market their properties in order to achieve a number of outcomes. Firstly, there was a need to market the destination of the Yasawas which branding was lacking. Everybody was going out and doing their own thing. Secondly, the tourism operators simply didn't communicate, nor did they achieve strength through numbers in providoring or advocacy with the authorities. As a result, they were paying retail for many of their goods, and didn't have any relationships at the Ministry level. They all operated without licences.

Thirdly, he identified the need to focus the operators on better tourism practices. Richard gathered together all of the players and formed the Nacula Tikina Tourism Association. After the first few meetings, he challenged them to create a responsible tourism charter, which they did and have now all adopted, covering such things as waste management, Fijian culture, protection of ecology and environment, guest safety and care, education and training, communication, and self improvement. In fact, they think sufficiently highly of it now that it is published in their small brochure! As a result of this initiative, and this commitment to environmental excellence and responsible tourism behaviour by the properties, occupancy in the region has increased to the point where one of the large tourism transport operators in Fiji is now running a daily boat with capacity for 260 guests to the region. 12 months ago, there was limited transportation for only 60 people, 4 days a week.

National and Regional Planning Strategy

Every national and regional planning strategy should strive to improve the social and economic levels of human communities while maintaining environmental integrity in perpetuity. In other words, planning should contribute in an important way to sustainable development, which comprises all human activities. For this reason we should ensure that tourism, which has attained a major socio-economic role around the world, is balanced with other economic, social and environmental objectives.

A national tourism strategy should be firmly based on profound knowledge and wise use of environmental resources, which includes biodiversity. This will be achieved only if we foster harmonisation among all public policies, including tourism and biodiversity conservation planning. This harmonisation must occur at the national level, as well as at the regional (sub-national) and local levels.

Both tourism planning and biodiversity conservation planning lend themselves to a bio-regional approach. It is important to promote valuing of natural resources and tourism assets for inclusion in national accounts. The importance of tourism has to be promoted at a national scale.

One of the challenges that governmental tourism agencies face is that usually they are low priority compared to other government bodies. Tourism agencies need to do a much better job of understanding and demonstrating the importance of tourism to national and regional economies. "Use-it-or-lose-it" conservation strategies aim to achieve their goals through strictly controlled access to ecologically significant areas and resources - and tourism is a key component in this approach, giving value to the natural resources, including biodiversity.

Since there is the need for communities to appropriately value their resources, local education in the environmental and tourism fields are urgently required. It is very important to recognise the need for a symbiotic relationship between biodiversity conservation planning and tourism planning.

Most countries around the world still do not dedicate enough effort towards the conservation of biodiversity, perhaps because they haven't recognised the value and usefulness of this effort. Keeping the biodiversity resource base means enhancing the tourism attractiveness of a country. Unfortunately, in many developing countries, any change proposed in favour of an improvement in the conservation of biodiversity is interpreted as an obstacle for development.

The sustainable positive impact of biodiversity within a comprehensive regional development strategy that considers overall qualitative benefits and not merely quantitative economic gains within a partial sectoral focus is still not understood by the population at large, nor by many politicians. It is important to recognise that many rural communities located in some of the most attractive ecotourism destinations are characterised by extreme poverty.

Many government development plans don't consider their involvement in tourism activities, especially ecotourism, as an option to alleviate poverty.

Thus, it is very important to insist on promoting ecotourism to governments as an important tool for rural poverty alleviation programmes. There is a generalised lack of information on the environmental impact of the different economic activities on biodiversity and of mechanisms to evaluate and monitor this impact.

The key factor for success for Costa Rica's tourism sector is sustainability as much of tourism activity as of natural resources. The State plays the role of coordinating entity and regulator of projects and programmes that provide incentives to the communities, and also promotes and generates a real need for a sustainable model as part of environmental, business and local participation schemes.

Tourism activity is one of the most important ways of valuing biodiversity and backing its conservation. National and foreign tourists are counted among the National System of Conservation Areas (SINAC's) main clients. It is they who currently generate the greatest amount of resources for the institution.

With the aim of using tourism as a positive tool for the management of the protected wilderness areas (PWAs), SINAC has defined as its general policy facilitating sustainable tourism development based on responsible practices of planning and management that are in accord with actions for conserving the country's natural and cultural heritage.

Tourism in the PWAs seeks to prevent environmental damage, foster the satisfaction of visitors, provide support to the monitoring of sustainable tourism and is interested in contributing to the country's local economies. This is a good example of a clear objective in planning to use tourism as a tool for natural resource management.

In South Africa, taxing pollution and subsidising products and activities with less environmental damage (including ecotourism), as well as altering interest rates to encourage certain land use patterns are economic mechanisms that are efficiently being used by the government.

In Mexico, the Ministry of Tourism (SECTUR), in coordination with the Ministry of the Environment (previously SEMARNAP, now SEMARNAT), the Commission for the Knowledge and Use of Biodiversity (CONABIO) and several other institutions from the public, private, social and academic sectors, published in the year 2000 a National Policy and Strategy for Sustainable Tourism, which represents a good inter-sectoral effort and contains valuable guidelines and action plans.

In UK, a good example of striving to alleviate poverty is the British government's Department for International Development (DFID's) pro-poor concerns and their interest in funding tourism. Establish a national tourism strategy that prominently includes guidelines for biodiversity conservation planning. For conserving biodiversity through sustainable tourism:

— provide a strong scientific basis,
— adopt an integrated management approach that covers all socio-economic aspects of a region, including tourism,
— carry out ongoing monitoring of environmental impacts, through effective application of Environmental Impact Assessment (EIA), which should be carried out by inter-sectoral technical entities,
— apply the precautionary principle.

Promote and strengthen the decision making process and the standardisation (norms) process in a participatory manner, especially at the local level. Build up institutional capacity of the environmental authorities for follow up of environmental impact assessment and application of prevailing norms. Apply integrated land use planning at a regional scale, taking into consideration the local communities' opinions.

Within the framework of the national tourism strategy, establish a sound ecotourism policy of an inter-sectoral nature that will provide viable options for biodiversity conservation and sustainable development especially at the local rural level.

Concentrate efforts on mid and long-term development policies without being pressured by periodical changes in government administration. Foster the use of taxes, subsidies and interest rates to reduce the negative environmental impacts of economic activity and enhance the positive effects. Making loans more easily available is a key point, in which the government should also participate.

Both biodiversity conservation planning and tourism planning and development are complex, interdisciplinary and inter-sectoral phenomena, thus appropriate interaction and integration is even more so challenging. There is an overall need to develop effective inter-sectoral mechanisms that will ensure the harmonious interaction among all stakeholders and a symbiotic linkage between biodiversity conservation planning and tourism planning and development.

For this reason, practical and dynamic coordination bodies (whether they be called commissions, committees, councils or agencies) for ensuring this appropriate interaction must be set up. These bodies should include genuine, fully empowered representatives of the different stakeholders or sectors: government, private tourism sector, NGOs, local communities, and financial institutions.

Although it is difficult to have the tourists themselves in a committee of this sort, their input should be sought. Inter-sectoral cooperation for the appropriate, sustainable and interactive development of tourism and biodiversity conservation planning must occur at the local, national, and international levels. Jurisdictions and responsibilities of the different agencies, authorities and organisations should be clearly defined and complement each other.

Government representatives in these inter-sectoral coordination committees should at least come from the ministries (or equivalent government or para-statal agencies) dealing with tourism and the environment, and it is convenient also that the ministries dealing with agriculture, fisheries, and education be included. Sometimes it is not necessary to create a new entity.

A preexisting body may be strengthened so as to develop a mandate of cooperation and non-duplication - one which is legislated, if need be. In a number of countries, these inter-sectoral coordination bodies have been established, with varying degrees of success. In every case it is important to first create the appropriate political environment between government departments (tourism, environment, agriculture) before bringing the other sectors (NGOs, tour operators, local communities) together.

The role of NGOs is vital, and should include local, national and international NGOs. Bilateral and multilateral development aid agencies (where appropriate) should collaborate, but always serving the local and national interests of the corresponding country. There must also be environmental education for those who are taking decisions, both in the fields of conservation and tourism.

In Australia, CRC is a network of Cooperative Research Centres in Australia, one group of which is dedicated to sustainable tourism. The Sustainable Tourism CRC is a very effective model of how universities can work with government and the tourism sector to develop and implement a research agenda that works for all three groups.

Housed in universities throughout Australia, the research is undertaken by academics but directly responds to the needs of government and industry. It helps all facets of the sustainable tourism industry to become linked. In Chile, in 1994, an entity called the National Commission for the Environment (CONAMA) coordinated through the President's Secretariat was created.

In Costa Rica, this country has successfully integrated the management of national parks, wildlife and forests into a single organisation, the MINAE's (Ministry of the Environment and Energy) SINAC (National System of Conservation Areas). The system is made up of eleven Conservation Areas that comprise 25% of the entire country, providing a good scenario for inter-sectoral synergy. Strengthening of inter-sectoral coordination is achieved through clear mechanisms for participation of the private and state sectors for planning tourism: e.g., the National Accreditation Commission for the Certification for Sustainable Tourism (CST).

In Costa Rica, communities have formed associations and cooperatives in different places throughout the country that have, as their main activity, local tourism. Private reserves, which at the moment total more than 100, and cover 1% of the national territory, rely on ecotourism as their main source of revenue. They carry out their activities in coordination with, and backing from, the Costa Rican Tourism Institute (ICT) and the MINAE, in the case of ecotourism, and in the case of agrotourism projects, with the ICT and Agrarian Development Institute (IDA).

In Costa Rica, the Bandera Azul Ecológica ("Blue Ecological Flag") programme was established by ICT in 1996 by Executive Decree as a coordinated action carried out by MINAE, AyA (Institute of Water and Sewage), CANATUR (National Chamber of Tourism), and the Ministry of Health. This programme awards the Blue Flag to Costa Rican beaches that comply with requirements of cleanliness, environmental education, and community organisation, thus warranting the sanitary and aesthetic qualities of the beaches.

The Belize Tourism Industry Association (BTIA) was formed in 1985 to bring together tourism-related interests to meet the challenges of the industry and act as an important link between the public and private sectors. BTIA has chapters in each district of Belize, and includes a broad range of individual members from all areas of the industry as well as representation from associations such as the Belize Hotels Association, Belize Tour Guides

Association, Belize Tour Operators Association, Belize Ecotourism Association and others.

A perceived weakness of BTIA is that it functions more like a membership-based organisation rather than an umbrella organisation. As the latter, it would be able to function more directly as an advocacy for the industry. In Canada, there are a number of senior level government committees which are designed to enable united action across the country, in the various jurisdictions which have independent authority over resources.

In Canada, the Canadian Tourism Commission (CTC) developed their vision and mission tapping into the direction of a 20-member team of industry experts, then presented them to industry stakeholders, provincial and territorial governments, and destination marketing organisations, to get consensus on a shared vision and mission, and to encourage a unified Canada perspective.

In Canada, the Biosphere Reserve Ecotourism Initiative is one of a number of tourism partnerships in the CTC Product Club Programme. The Ecotourism Product Club is a partnering programme between individual Canadian Biosphere Reserves, local communities and the private sector. The goals of this partnership are to: encourage and package sustainable tourism opportunities based in nearby communities and tapping into the resources of the Biosphere Reserves; and help communities value protected areas through demonstrating that economic benefits can emerge from ecotourism.

Strengthen the coordination of tourism and biodiversity conservation planning government policies at both the national and local levels. Foster the participation of biodiversity conservation planners in meetings dealing with tourism planning and, conversely, the participation of tourism planners in discussions of biodiversity conservation issues.

Tourism needs to be involved in biodiversity and biodiversity in tourism. Create dynamic and practical inter-sectoral mechanisms (e.g. committees) for effectively coordinating interaction between biodiversity conservation planning and tourism planning and development. Ensure that these committees have representatives not only from the government sector, but also from the tourism industry, NGOs, local communities and the universities.

Nations around the world have growing economic needs and generally growing populations. However, their land area normally remains the same and this produces many conflicts regarding use of the land and its resources. Utilisation of land for tourism is becoming an increasing trend in many countries. It is important to ensure that national development plans contain a set of development guidelines for the sustainable use of land, water and natural resources.

The development of a diverse tourism base that is appropriately integrated with other local economic activities should be promoted. The only type of tourism which should be permitted in ecologically relevant and fragile natural areas is ecotourism. Ecotourism provides viable economic alternatives for other activities which are less sustainable—and many times downright destructive to the environment—, such as uncontrolled logging, extensive animal husbandry, mass tourism or mining.

This may bring about conflicting interests of water supply and other vital resources between tourists, the needs of the local population, and intensive agriculture. For this reason effective national zoning plans for land use are urgently required (and of course, effective enforcement of those plans). Ecotourism should not be seen as a panacea or as a monoculture in a given area, but should be considered a complement for other sustainable economic activities that make wise use of the natural resources.

All relevant stakeholders should be involved in the development of sound management plans for land-use. Means for providing the organisation, facilities and enforcement capacity required for effective implementation of those management plans should be put in place.

In Australia, in the Shark Bay Region of Western Australia, the 1995 Shark Bay Regional Plan coordinated a World Heritage Strategic Plan, several tourism strategies and several fisheries management strategies. In order to effectively manage future development and use, the Department of Conservation and the Ministry of Planning worked together to ensure that their planning was complementary and there were no areas of conflict as regards use of land and coastal zones.

In Costa Rica, the Conservation Area is an original national regionalisation structure created by MINAE (Ministry of Environment and Energy) and SINAC (Conservation Areas National System). A Conservation Area is defined as that territorial unit governed by one and the same strategy for development and administration, where there is interaction between private activities (including tourism) and state activities on issues of management and conservation of natural resources, and where sustainable development solutions are pursued jointly with civil society.

In this way, SINAC's administration covers all of the national territory (including protected areas) where exploitation of biodiversity activities and natural resources in general are promoted and regulated. Each area has a set number of protected wilderness areas (PWAs), including buffer zones. There is coordination between state institutions and civil society that carry out activities in the areas.

In Costa Rica, as for protected areas, SINAC in coordination with NGOs, has widely invested in the promotion of the different and numerous PWAs in the country and in bettering the infrastructure for attention to the public. This effort has had its positive impact, as was shown by a recent survey on tourists ' perception on international tourism but more so and especially on national tourism, as reflected by the growing number of visitors to the PWAs.

In Costa Rica, both in conservation areas and protected areas the ENB (National Biodiversity Strategy) establishes a line of action and top priority activities according to five-year planning periods, where the tourism sector is a key factor. In the Strategy, ecotourism with ample participation by civil society and in coordination with the government, is visualised as the top priority tourist activity to be developed and strengthened for conservation and sustainable use of biodiversity.

Ensure that tourism planning is undertaken as part of the overall development plan for any area (including its land-use plan), and not undertaken in isolation. Focus on ways in which different interests can complement each other within a balanced programme for sustainable development, enabling the different stakeholders to work alongside each other.

Minimise negative impacts of tourism on the natural and cultural environment and promote tourism as a tool to protect important natural habitats and conserve biodiversity in accordance with the Convention on Biological Diversity. Ensure that, wherever tourism occurs or is liable to occur, the corresponding land-use plan should include a component of tourism land-use, carefully zoning the areas as regards the type of tourism that should take place: high, middle or low intensity.

The inclusion of natural areas (and provisions for their conservation) is a vital element of any zoning plan. Ensure that the only type of tourism that will take place in vulnerable and fragile natural ecosystems follows the principles of ecotourism, considering it a viable option for minimising negative impacts and promoting positive environmental and socio-economic contributions.

Ecotourism should always be carried out with the active involvement of the local communities. Foster the creation of links between natural protected areas and other ecotourism destinations by means of biological corridors that will amplify biodiversity conservation to a larger regional level.

Sustainable Tourism Industry

In every case, tourism should pay attention to the "triple bottom line": economic, environmental and social factors must be attended to simultaneously. This implies the need for integrated management and the adoption of an ecosystem approach, as advocated by the CBD. Continuous management of tourism is just as important as proper planning and development. It is imperative to provide incentives for the wide range application of environmental management systems.

The only viable relationship between tourism and nature conservation is a symbiotic one. It is not enough to have a situation of coexistence and certainly nobody benefits from a conflictive relationship. Tourism management needs to form part of biodiversity management planning. The allocation of land uses must be carefully coordinated and inappropriate activities that damage ecosystems should be strictly regulated.

This may be done only by strengthening and developing integrated policies and management that cover all socio-economic activities in the different ecosystems, including terrestrial, coastal and marine zones. Management solutions are also needed for simple, but persistent, problems such as litter. It must be emphasised that enjoyment of biodiversity and natural areas is not only for rich foreigners, but for all national inhabitants.

Ecotourism is made up of visitation by both national and international tourists. The former component is usually more sustainable than the latter if a sufficient standard of living exists in the country (i.e. domestic tourists possess the financial means to visit, and consequently support, protected areas). The different sectors must understand the tourism market for cultural and natural heritage products, and how this is linked to tourism's ability to support conservation through product demand.

Understanding the experiences and products tourists are looking for, enables protected area managers to tailor certain aspects of the destination for the desired type of tourist. Accurately forecasting the amount of anticipated visitors enables planners to lobby for and develop sufficient infrastructure.

It is important to demonstrate how the private sector can implement environmental management plans, using low cost methods first, and then use any left over money to retrofit, making the tourism facility more sustainable. It is necessary to show the large hotel chains that environmental management brings a profit. Using environmentally friendly techniques saves money for hotels and all other tourism service providers.

In Australia, a national agenda to support sustainable tourism (including ecotourism) as a tool for conserving biodiversity and for better use of natural areas. The National Tourism Strategy was formulated in 1992 to, among other goals, enhance community awareness of the economic, environmental and cultural significance of tourism.

The Strategy's environmental goal is to provide for sustainable tourism development by encouraging responsible planning and management practices consistent with the conservation of Australia's natural and cultural heritage. In 1994, within the framework of this strategy, the National Ecotourism Strategy was published also by the Department of Tourism.

In the strategy, integrated management of biodiversity resources and sustainable tourism, with ample participation of civil society and in coordination with the government, is considered as the top priority for national development. As part of sustainable tourism, Costa Rica seeks an activity where there is better distribution of resources in the different regions of the country and direct involvement of rural communities, while at the same time minimising negative environmental impacts.

In Cuba, following the 1992 Earth Summit, where Cuba was one of only two countries to obtain the highest rating for implementing sustainable development practices. The government set up a National Programme for Environment and Development and created a series of new institutions to continue along a course of sustainable development. The new institutions include a National Commission on Ecotourism, made up of tourism officials, environmentalists and scientists, created to ensure integrated management of biodiversity resources and tourism activities.

Conventional Mass Tourism

The conventional mass tourism is still the mainstream of the tourism industry and it is quite probable that this situation will prevail for some time. For this reason it is vitally important to aim our attention on mass tourism, striving to apply measures to make it more environmentally friendly and minimising its negative impacts on biodiversity. We should not consider only ecotourism linkages with biodiversity conservation, but also linkages of mass tourism, especially the effects of big hotels on the environment and how their design and operation can become more environmentally friendly.

At a global scale, perhaps providing a number of ecolodges is not going to make much of a difference- ultimately we have to affect the larger tourism industry. This means we have to consider how to improve the environmental record of very different items like airlines, airports, big amusement and theme parks, golf courses, and sports stadia.

Training to develop skills of hotel owners and operators to understand what sustainable tourism is and education about best practices are vital activities. There is a need to strengthen and to revise legislation so that this approach is well understood and widely disseminated.

Environmental legislation should act as a motivation force, and also as a base for certification. Also, a widespread educational campaign so that tourists will be demanding environmentally-friendly hotels is urgently needed. Tourism shouldn't be only market driven. In Africa, for example, people feel bad about tourism use proscribed to the community.

A cause for conflict arises when developing nations are told to be sustainable whereas western countries can have the huge hotels. It is vital to disseminate codes of ethics for conventional tourists, which will serve as a tool for alleviation of negative impacts. The effects of negative impacts are frequently long term and not always obvious in the short term.

Saving water and energy by reducing the number of towels used in hotel rooms has become a cliché - but only because notices have made a difference in hotels around the world. In analysing mass tourism impacts, both new tourism facilities and preexisting tourism facilities must be considered. In the former case, the application of minimal environmental standards for siting of new tourism services and facilities is urgently required.

In the latter case, methods for improving the operation, making it more environmentally-friendly, should be applied, through retrofitting or adding new, more appropriate technologies. In every case, the benefits to the tourism sector (market demand, economics, effective management) must be persuasively demonstrated. It is not a matter of sanctions and pressuring, rather encouraging the tourism sector to become more environmentally friendly (which will result in economic benefits for them).

For example, water heating in many conventional hotels around the world is currently very inefficient and costly, so that wide use of alternative energy sources should be more than welcome by mainstream tourism operations. Also, many traditional beach destinations are experiencing a loss of repeat visitors because of water pollution, so that more environmentally-friendly practices are definitely in the interest of beach resort owners and operators.

Cruise ships cause enormous environmental damage. It is estimated that they discard many thousands of tons of untreated waste into the oceans of the world every day. Strict regulations have to be applied to this type of destructive tourism.

In Australia, for the 2000 Olympic Games the Sydney Organising Committee for the Olympic Games (SOCOG's) mission was "to deliver the most harmonious, athlete oriented, technically excellent and culturally enhancing Olympic Games of the modern era". SOCOG used its best endeavours to set a new standard of environmental excellence for organising and staging a large sporting event. To this end, SOCOG was committed to strict environmental guidelines, guided by the principles of Ecologically Sustainable Development (ESD).

Environment was considered by the organisers as the "third pillar of Olympism". Among the major environmental achievements were: the SOCOG Environment Programme was established in 1996, early enough to allow sufficient time for staff to be integrally involved in planning; carrying out of the Olympic Greenhouse Challenge, a major project to assess the greenhouse impact of the Games for minimising greenhouse gas emissions; a programme of environmental education (including a waste education plan) as a component of staff training; an environmental specification for sponsors, licensees and suppliers; an integrated waste management solution; packaging and foodware specification to control inputs into the waste stream; an Olympic Results Information Service (ORIS), an electronic system which reduced the huge amount of paper required to provide media with results; saving water and energy at all Olympic buildings and facilities; facilitating access of spectators through public transport, resulting in energy conservation and pollution and greenhouse gas avoidance; Olympic merchandise had minimal packaging and minimised the use of PVC.

Encourage collaborative research on the impacts, promotion and incentives of mass tourism. Enforce current environmental laws, regulations and norms on waste management, air pollution and monitoring devices. Overcrowding, misuse of natural resources, polluting of air and water, the construction of infrastructure and facilities, and other activities associated with tourism, all produce impacts on the environment. These impacts may be not only physical, but also cultural.

Negative impacts of tourism vary according to the nature and number of tourists, the type of physical facilities and the way tourism is managed. The individual tourist normally has a relatively small impact. Problems arise, however, if the number of tourists is large or the resource overused.

Thus although tourism can be a lucrative source of revenue for a country, a tour operator or a specific destination, it can also represent a cause of serious damage to the environment, including the biodiversity resources. Tourism impacts on the environment are manifold: impacts on geological exposures, minerals and fossils; on soils; on air and water resources; on vegetation; on animal life; on sanitation; on the cultural environment; and aesthetic impacts on the landscape.

These different impacts, which are actually manifestations of change on the environment, rarely occur singly and their ecological effects are usually very complex. There are different ways of minimising negative impacts of tourism: through widespread environmental education of tourists and the tourism industry, through strict enforcement of laws and regulations, and through effective visitor management.

Visitor management begins before arrival, then occurs on the ground at the site, and finally after the visitors leave through continual communication. The process of environmental impact assessment (EIA) is one of the most effective methods for determining whether a project will be sustainable, and if so, for developing safeguards to ensure its continuing sustainability.

EIA aims to ensure that the likely outcomes of any development (including tourism developments) are addressed at an early stage so that disastrous environmental and social consequences can be avoided. EIA applied to tourism projects should play a crucial role in government-decision making in every country.

Zoning, a vital component of land-use planning, is the process of applying different management objectives and regulations to different parts or zones of a specific area. Zoning is a tool required in any land or water-use plan (including marine, coastal and fresh water areas) and should always be used to regulate utilisation of the land and water resource over the long term.

Zoning should be applied at all the different levels: national, regional and local. In the first case, zoning plans should be carried out by government authorities at the highest (e.g. federal) level. In every case, zoning must be comprehensive, considering the different socio-economic activities (obviously including tourism, when it occurs) , but also natural areas which should be left undeveloped.

Zoning plans should solve conflicts of interest with water supply and other vital resources, including biodiversity. Understand that some areas of relevant biodiversity should be conserved for their own value and that not all relevant biodiversity zones should experience visitation. Biodiversity has is own value.

As regards tourism, it is important to have a zoning scheme which should cover the different possible tourism activities. A good example of a comprehensive tourism zoning plan includes the following specific zones:

a) Strictly protected zone, where the presence of all types of tourists and tourist infrastructure are strictly prohibited.

b) Restricted tourism zone, where access is allowed only to a limited number of tourists, usually on foot.

c) Moderate tourism zone, where visitors are encouraged to carry out diverse low -impact activities compatible with the natural and/or cultural environment.

d) Semi-intensive tourism development zone, which should always be an area of limited extent (especially when near environmentally-sensitive natural areas), where some moderate-impact facilities are included (e.g., ecolodge, visitor centre, limited parking areas).

e) Intensive tourism development zone, which should only occur in highly popular mass tourism destinations (e.g. beach resorts, ski resorts, amusement and theme parks), where a considerable degree of concentration of tourists and tourist facilities take place.

Obviously, in ecotourism destinations (especially in protected areas), there is nor room for this zoning category. But, even in the case of mass tourism destination areas, planning should endeavour to minimise negative impacts, including pollution of air, water, and soil resources. The different zones indicate where and what type of physical infrastructure and services should be provided, by means of a clear categorisation of modality and intensity of land use (and use of natural water resources and other natural resources), striving in every case to minimise negative impacts on the natural and cultural environment, as well as optimising the ecotourists' experience.

Zoning also indicates to us where facilities, activities or services should not be developed. In essence, a zoning scheme shows the development suitability of the different portions of a site. Those activities which are carried out in each zone are normally mutually exclusive (and often conflictive), so that zoning decisions must be taken very carefully. Allow diversity through wise zoning of tourism activities, providing tools so that people (both locals and tourists) can decide upon which zone they fit into (or they prefer to visit) and then follow the regulations related to their zone.

Environmental Carrying Capacity

Environmental carrying capacity is the capacity of an ecosystem to support healthy organisms while maintaining its productivity, adaptability, and capability of renewal. Tourism carrying capacity is a specific type of environmental carrying capacity and refers to the carrying capacity of the biophysical and social environment with respect to tourism activity and development.

It represents the maximum level of visitor use and related infrastructure that an area can accommodate. If it is exceeded, deterioration of the area's environmental resources, diminished visitor satisfaction, and/or adverse impacts upon the society, economy and culture of an area can be expected to ensue.

The basic components of tourism carrying capacity are:

— biophysical,

— socio-cultural,
— psychological, and
— managerial.

Over the last decade or so the tourism carrying capacity concept—and several related methodological tools—have been heavily criticised as being oriented excessively towards quantitative considerations. Alternative methodologies, such as LAC (Limits of Acceptable Change) and VIM (Visitor Impact Management) have been developed (especially for relatively undisturbed natural areas. The shift in attention from an appropriate use level to the desired condition is the basis of LAC's revised approach to visitor carrying capacity.

The LAC approach concentrates on establishing measurable limits to human induced changes in the natural and social setting of a specific area, and on identifying appropriate management strategies to maintain and/or restore desired conditions. VIM, developed by the National Parks and Conservation Association of the USA, is a technique for assessing and managing the environmental and 'experiential' impacts of increasing numbers of visitors to natural areas.

VIM recognises that recreational impacts on the environment and the quality of the recreational experience are complex and influenced by factors other than use levels. Traditional carrying capacity methods as well as LAC and VIM techniques are all management tools for minimising negative environmental impacts.

Certification System

Every country needs a national certification system in the field of tourism; it needs to be applied by an inter-sectoral body (that should include NGOs, communities, tour operators, etc.). National governments need not be the certifiers. The certifier is preferably the result of a cooperative effort (e.g. a commission) that involves scientists, communities, industry, etc.

The certification programme must have official recognition by all stakeholders. Certification schemes need to be independent, non-profit, transparent, and credible. If certification is to be used in the field of sustainable tourism it must bring benefits to biodiversity conservation. A standard is a level that every one must meet (e.g. safety standards for being a river raft guide); certification should place more requirements, raising the bar in terms of the criteria that need to be met.

Certification is something destinations must decide for themselves and should be carried out on specific measured environmental performance. Local or regional certification schemes are normally preferable, but based on international guidelines, and they must include biodiversity as one of the main components.

Certification should provide economic benefits to all stakeholders (including tour operators). Most tourism certification in the past has not duly considered factors like environmental quality, biodiversity conservation, and respect for local cultural traditions. In the future, certification should appropriately link tourism with biodiversity conservation, and should be performance-based. Biodiversity (including conservation, maintenance, and enhancement), socio-economic (comprising gender issues , level of participation of local people, and benefit sharing), and cultural criteria must be considered in every certification scheme.

It should be recognised that many certification programmes are currently unreliable (resulting , e.g. in lack of training in hotels). Try to attract tourists who are already complying with certification schemes. In Australia, the National Ecotourism Accreditation Program (NEAP) was launched by the Ecotourism Association of Australia in 1996, and is jointly run with the Australian Tour Operators Network.

The NEAP arose out of the fundamental problem of how to distinguish between genuine ecotourism operators and other operators who operate in natural areas. At the end of 1998, there were about 130 attraction, accommodation and tour products accredited, according to eight basic sustainability principles. Depending on how many points operators achieve, they can either be awarded accreditation or advanced ecotourism accreditation.

The Accreditation Program has been critical in helping ecotourism operators improve the profile of their products, which in turn has led to greater customer recognition and an emerging market edge.

The Mohonk Agreement is an agreed framework and principles for the certification of ecotourism and sustainable tourism, which was unanimously adopted at the conclusion of an international workshop held at Mohonk Mountain House, New Paltz, New York in November 2000. This agreement contains a set of general principles and elements that, according to the workshop participants, should be part of any sound ecotourism and sustainable tourism certification programmes.

Participants came from 20 countries and delegates represented most of the leading global, regional national, and sub-national sustainable tourism and ecotourism certification programs, conservation and environmental organisations, and others with expertise in tourism and ecotourism certification and environmental management. Workshop participants recognised that tourism certification programmes need to be tailored to fit particular geographical reasons and sectors of the tourism industry, but agreed on a basic set of principles that, in their opinion, must frame any ecotourism and sustainable certification programme.

Education and Training

Education and training are vital for enhancing the links between biodiversity conservation planning and tourism. Appropriate environmental education and awareness-building among local communities, the private sector, tourists and government will help to promote responsible actions to reduce adverse impacts and increase support for conservation policies and programmes.

Frequently, negative social and ecological impacts caused by tourists result from the fact that guests have little or no understanding of the local culture and ecology. It is very important to increase public awareness of environmental issues through tools such as TV and radio programmes, magazines and posters. Consumer education campaigns are needed to support this process.

Governments can further promote environmental and social goals through general awareness-building campaigns focused on a wide range of groups, including hotel owners, tour operators, government offices, tourists, school children, and local communities. Education for tourists should include specific examples of a region's biodiversity, its particular ecosystems and any endemic species present.

Education campaigns for the tourism sector should include an ethical mandate, in the sense that tourism needs to benefit both the natural and social environment, improving (and investing in) social and public infrastructure, as well as endeavouring to eliminate local poverty. Hotels, restaurants and tour operators are all encouraged to carry out ongoing training programmes among their staff, striving to teach them sound environmental practices.

The interaction of tourism and biodiversity conservation should be imparted at schools where tourism is taught (at different levels, i.e. technical, university, graduate). Innovative partnerships should be set up between the different sectors, so as to achieve a better interaction between biodiversity conservation planning and tourism activities.

Linking tourism revenue with conservation management should be promoted, as well as marrying different priorities of conservation authorities and private operators. Partnerships should always be mutually beneficial. It is important to apply methods such as incentives, compensation and certification to promote environmental best practices in tourism, including the use of mechanisms such as taxes or voluntary strategies for sponsoring reserves.

The World Bank and other multilateral agencies (Organisation of American States, Commission of the European Communities, South Pacific Commission, etc.) have produced many practical manuals for developing capacity building in various sectors, including tourism and environmental conservation.

Better information exchange must exist at all levels: between the different sectors, within these sectors, and among countries. Today's consumer must be encouraged to be more discriminating. The 21st century offers information and communications mechanisms which were not previously available only a few decades ago. The spread and progress of mass media such as TV, radio, newspapers, specialised magazines, and especially the Internet provide extraordinary tools for disseminating information linking tourism with biodiversity.

Other new technologies like the use of CD-ROM open enormous possibilities. Traditional tourism information should be revamped, so as to include clear information on biodiversity, natural ecosystems, endemic species, conservation issues, cultural traditions and attractions. Information must be scientifically precise, but presented in an uncomplicated and attractive way.

From brochures to information offered by tour guides, points of contact offer key chances to create more considerate and responsive clients. However, in spite of all the technological breakthrough in information and communications systems, it is still believed that over half of all tourist choices are based on word-of -mouth, so it is important not to raise expectations on site that cannot be met.

Many countries already have a long standing tourism reputation as conventional tourism destinations, but the international public usually retains the standard images, so that these have to change to reflect new images, including nature. In the case of Mexico, traditional images of mariachi musicians and beautiful beaches should be complemented by images reflecting the enormous biodiversity richness of that country.

Tourism is a big industry based on many small businesses. If we want tourism to contribute to biodiversity conservation and sustainable development, tourism must then be a profitable enterprise, providing substantial economic benefits to the people involved in the different facets of this complex activity. Even in the case of ecotourism or rural community tourism, the process must be characterised by being a good business.

While significant major players offer large numbers of popular packages, even in mass destinations the small businesses are the bedrock of visitors' experiences. The importance of small businesses to nature tourism in rural areas of developing countries cannot be overstated. A code of ethics based on "Profit with principles" should be present in every tourism business. Especially in the case of ecotourism and rural community tourism, profits should be analysed on a long term basis. Aiming for quick profit may bring catastrophic environmental and cultural results.

More appropriate funding mechanisms are required in the tourism industry. Unfortunately, there are no specialised banks for tourist investment, nor are the banks usually aware of the sector's specific needs. However, there are a number of financial mechanisms which may be applied for funding tourism projects, such as venture capital schemes and through the private equity market.

In the case of new technologies, all interested credit cards need to work to assist the tourism sector (including local) communities to enter into e-business as regards ecotourism and other community-based enterprises. Governments are often focusing on larger products to the disadvantage of all small tourism businesses. Community products may be further disadvantaged by their remote location and isolation from other products.

Micro-credit programmes should be set up for developing tourism in rural areas. Financial training is widely required for successfully launching micro-credit initiatives. As regards funding for national parks and other protected areas, it is vital to have tourism participating in the provision of proceeds towards the conservation of these areas and the biodiversity therein contained. The appropriate pricing of entrance fees to protected areas is a key factor for channelling tourism revenues to the management and protection of these areas.

In order to promote tourism to environmentally significant destinations, both domestic and foreign tourists should be targeted. Domestic tourism is sometimes as much as ten times bigger than foreign tourism and by promoting nationals to visit their protected areas we will be fostering environmental awareness and pride related to the natural heritage of a country.

In many countries, frequently nature tour operators are left out of the directory pages of the standard tourism industry. Sustainable tourism marketing may involve the need for specific strategies like demarketing and niche marketing. In many African countries, for example, local people feel bad about tourism use proscribed to the community. Tourism is widely seen in these countries as an activity restricted to rich foreigners.

The International Finance Corporation (IFC), which is part of the World Bank Group, has a unique mandate: to encourage private investment, foreign as well as domestic, in developing countries. Founded in 1956 by 31 countries, IFC today has more than 160 member countries. It is the largest multilateral source of financing for private sector projects in developing countries, with annual investment approvals in excess of US$2 billion.

IFC also helps companies to raise capital in the international financial markets and provides a full array of advisory services to private businesses and to governments. IFC recognises tourism's contribution to economic development primarily through the support of hotel investments, although private sector tourism infrastructure can also be financed. IFC has provided loan and equity financing for hotels and other tourism businesses for over three decades.

IFC's involvement in the tourism sector focuses on projects that promote sustainable development, enhance conservation of natural resources and the environment, and promote high standards of environmental safety.

Promote conservation-oriented tourism, including ecotourism, through mass media and integrate into conventional mass tourism promotion and marketing. Target both national and international tourists. Develop opportunities through tourism shows to demonstrate and promote community tourism experiences and for communities to share experiences.

References

Cooper, Chris; et al (2005). *Tourism: Principles and Practice* (3rd ed.). Harlow: Pearson Education.

Isaacs, J.C. (2000). *The limited potential of ecotourism to contribute to wildlife conservation*. The Ecologist. pp. 28(1):61–69.

McLaren, D. (1998). *Rethinking tourism and ecotravel: the paving of paradise and what you can do to stop it*. West Hartford, Connecticut, USA: Kamarian Press.

Singh, L. K. (2008). Issues in Tourism Industry. *Fundamental of Tourism and Travel*. Delhi: Isha Books.

Vivanco, L. (2002). *Ecotourism, Paradise lost—A Thai case study*. The Ecologist. pp. 32(2):28–30.

7

Systematic Planning for the Conservation of Biodiversity

Biodiversity literally refers to the variety, distinctiveness and abundance of life forms and processes. It is within this literal context that Kuhnian normal science operates; e.g. ecologists' long intrigue with and debate over the role of biodiversity in ecosystems and biologists' systematic classification of taxa. Figuratively, biodiversity has become one of the most recognized environmental slogans of the decade. The notion of biodiversity moved beyond the comfortable realm of science following the global media coverage of the 1992 United Nations Conference on Environment and Development – the 'Earth Summit' – in Brazil. Joining the ranks of sustainable development, ecosystem health and other environmental heralds, biodiversity has become more than a literal scientific description of environmental phenomena and has entered the murky context of political and social debate. Biodiversity must now be considered from both a literal scientific perspective and a figurative political perspective.

With this transition, biodiversity has necessarily become an issue of ethics and ensuing values, beyond its traditional ecological roots. Furthermore, the traditional perspective of biodiversity is being challenged fundamentally by the emergence of a postnormal, systemsbased science. This newperspective, the hallmarks of which are uncertainty and complexity, reveals ecological realities that force consideration of broader perspectives and multiple contexts for biodiversity. Thus, the tools for biodiversity conservation planning are no longer limited to those of normal science but include those of systemsbased, postnormal science.

What are the implications of these new biodiversity perspectives and contexts? This central question is explored: first, through identifying emerging perspectives of biodiversity; and second, through tracing the emergence of a newecological context for biodiversity conservation. From this analysis, the challenges and implications for conservation decisionmaking are considered, and a systems-based approach to conservation planning and management is proposed.

Biodiversity conservation is a normative endeavour. Like medicine, the science that is used to support biodiversity conservation is motivated by an ethic; indeed it is almost analogous in the desire to 'heal and cure' pathologies. Consequently, in the conventional sense, there can be no strictly scientific 'ecological' basis for conservation16. Rather, there exists a range of valueorientations resulting from ethics that underlie conservation motivation.Interestingly, the current literature makes much of the 'scientific values' for biodiversity – a particularly revealing oxymoron where conventional Newtonian science is concerned. However, it is possible to consider this a 'trespass of semantics' for the moment because ecology is divided between a reductionist and systemsbased approach.

Systematic Conservation Planning

Systematic conservation planning is an effective way to seek and identify efficient and effective types of reserve design to capture or sustain the highest priority biodiversity values and to work with communities in support of local ecosystems. Margules and Pressey identify six interlinked stages in the systematic planning approach:

— Compile data on the biodiversity of the planning region
— Identify conservation goals for the planning region
— Review existing conservation areas
— Select additional conservation areas
— Implement conservation actions
— Maintain the required values of conservation areas

Conservation biologists regularly prepare detailed conservation plans for grant proposals or to effectively coordinate their plan of action and to identify best management practices. Systematic strategies generally employ the services of Geographic Information Systems to assist in the decision making process.

Conservation Planning Tools

In the 1980s, planning attempts were replaced by efforts to use detailed biogeographic distributional information for the design of conservation area networks and by an explicit attempt to achieve spatial economy within them (that is, achieve the biodiversity goals in as little area as possible). By the 1990s, efforts were also under way to include socioeconomic criteria in these designs. These new approaches were algorithmic and relied heavily on computers because they needed to process large quantities of data rapidly. The field of systematic conservation planning emerged from these efforts. Several protocols exist for systematic conservation planning.

These involve a number of stages including the determination of stakeholders in the planning region, collection and treatment of biological and socioeconomic data, selection of features to represent biodiversity quantitatively, selection of individual conservation areas (the problem studied since the 1980s), assessing vulnerability and the prognosis for components of biodiversity, and multicriteria analysis to satisfy divergent socioeconomic and biological goals of the stakeholders. Each of these stages is aided by the use of planning tools consisting of software packages implementing a variety of algorithms for these purposes. Besides researchers, these tools are typically used by intergovernmental groups, governmental planning departments, nongovernmental conservation organisations, and, on occasion, by groups of landowners who want to manage relatively large areas for conservation and other uses.

The past 20 years have seen the emergence of a wide variety of such software packages. The rapid growth in the application of these tools has not been accompanied by reviews to identify their differences and similarities or their relative strengths and weaknesses. Here, we review the concepts and techniques on which conservation planning tools are based and identify issues that future developers must confront. A forthcoming review describes the use of these tools by planners.

A conservation planning tool is defined as software with the following two characteristics:

— It can be used to guide decisions about conservation action for biodiversity, although it may also be used to plan for other natural values such as scenery or ecosystem services.

— At the very minimum, it can identify either (a) sets of complementary sites needed to achieve quantitative targets for biodiversity features or (b) the complementary contribution that individual sites make to biodiversity conservation within a region.

This definition is intended to bound the scope of this review to software tools designed to help planners decide on the location and configuration of conservation areas. Thus, it does not consider geographical information systems (GIS), software for population or habitat-based viability analysis, or niche-modelling software, which are routinely used in conjunction with conservation planning tools. Biodiversity is construed here to include the variety of living features and processes at all levels of structural, taxonomic, and functional organisation but not ecosystem services or culture-based categories.

The central goals of a conservation plan are *representation* and *persistence*. Representation requires that all relevant features of biodiversity are adequately accounted for in a plan. However, representation must be achieved with *economy*; that is, adequate coverage must be achieved at a minimum cost. Economy is also called efficiency in the literature, but here it is used to denote computational efficiency. Economy typically refers to spatial economy because the most relevant cost is usually that of designating land for conservation. Spatial economy is critical because conservation competes with other potential uses of the land; a conservation plan that ignores these demands has a reduced chance of implementation. In practice, owing to such constraints, entire conservation area networks have rarely been implemented. One exception is the set of areas recommended by Kirkpatrick in Tasmania. However, planning tools increasingly influence both policy and implementation.

The second central goal of conservation planning, persistence, refers to the need for planning to go beyond the representation of biodiversity patterns. If biodiversity is to persist in conservation area networks for centuries or millennia, then a variety of ecological and evolutionary processes must be accommodated. These processes include dispersal, local extinctions and recolonisations, species interactions, migration, patch dynamics, adjustment of species' distributions to climate change, and diversification of lineages.

The aim of conservation planning tools is to help users make decisions, not to exclude them from decision making. Planning tools are decision support systems not sensu stricto decision-making systems. The use of planning tools has been criticized on the grounds that they do not incorporate local expertise, require inordinate amounts of data not normally available, and are expensive to apply, diverting resources that are better spent acquiring land. These unjustified criticisms are based on misconceptions about how tools should be used and what they have achieved. Conservation planning is a dynamic process in which these tools are supposed to aid decision makers in identifying good policy options. This process is ideally carried out iteratively during a planning exercise, with results from tools guiding the delineation and refinement of policy alternatives. The rapid recent growth of biological and environmental databases has ensured that there is now no terrestrial region for which there is not enough data for planning tools to improve policy formulation. Several studies have shown that ad hoc policy formulation is remarkably cost ineffective in prioritising conservation areas.

In this review, the term "conservation area" is used instead of the more traditional "reserve" to indicate that reservation is one extreme of a continuum of policy options for management that can range from reservation to the restoration of degraded areas. The appropriate management regime depends on both the needs of the biota and sociopolitical constraints and opportunities.

Three key concepts have guided the design of planning tools: *complementarity, irreplaceability*, and *vulnerability*. Complementarity goes back to the origin of planning tools in the work of Kirkpatrick and coworkers. Their crucial insight was that, if the goal is to represent biodiversity maximally in a given land area, then sites should be selected to maximise the differences in their biotic content. This principle, later called complementarity, was independently discovered at least three other times in the 1980s. The term "complementarity" was introduced by Vane-Wright et al. The complementarity value of a site, relative to an existing set of prioritised sites, is its quantitative contribution to the representation of biodiversity features that are not adequately represented in the existing set; here adequate representation consists of meeting predefined targets. Differences between sites are critical to representing biodiversity adequately;

complementarity is thus related to β-diversity. However, complementarity is an essentially asymmetrical relationship between biotas. The use of complementarity requires information on the specific content of sites, namely, lists of surrogates present in them; summary statistics such as richness do not suffice. The complementarity value of a site must be updated whenever the set of prioritised sites changes.

A second key concept is irreplaceability. A site prioritisation problem typically has multiple solutions-some closer to optimal than others. In this context, although finding a single solution can provide an indicative answer about the size and configuration of the required set of sites, single sets are of limited value in practical planning for two reasons. First, some of the unselected sites might be useful replacements for a selected one that cannot be included in a conservation area network. Second, it is unclear whether any particular selected site is essential for achieving targets or whether it could be replaced by others and is therefore negotiable. These limitations may be addressed by determining the irreplaceability of each site across the planning region. This indicates the probability of a site being needed to achieve targets. Complementarity is implicit in irreplaceability. Irreplaceability can be measured exactly for very small data sets by exhaustive analysis of all possible site combinations and then determining the proportion of representative combinations achieving all targets that contain each site. A variety of techniques have been developed to estimate irreplaceability in more complicated contexts.

Because the persistence of biodiversity is a crucial goal of conservation planning, a third critical concept guiding the design of planning tools is vulnerability. It may be inappropriate to include a site in a conservation area network if its use becomes incompatible with management for biodiversity or if its biotic components have low abundance or probability of persistence. Vulnerability refers to both. In the planning tools that exist today, concern for vulnerability is incorporated mainly through the specification of conservation targets, preferences between sites when developing plans, and scheduling of conservation action on the ground during the implementation phase.

Design and application of conservation planning tools occur in a dynamic context in which plans must be continually updated on the basis of new information. Additional or improved biological or other data might become available after initial plans have been formulated. The preferences of planners might change as a response to either the new data or sociopolitical developments. Initial plans may become partly unachievable because of the loss or degradation of some selected areas. With these goals in mind, we use six criteria to judge the performance of planning tools: economy, efficiency, flexibility, transparency, genericity, and modularity.

Levels of Planning

Planning must take place, sometimes simultaneously, at a wide range of spatial scales from local plans (for areas of ~100 km^2) to those for entire continents. Typically, planning tools are not restricted to particular spatial scales. However, data availability varies between scales, biological processes differ from level to level, and conservation priorities and decision-making scenarios may also vary with scale for the same region. Methods and tools appropriate at one scale may not be appropriate at others. The advent of GIS technology has made the hierarchical representation of spatial features at different scales easier; nevertheless, this only partly mitigates this problem.

Equally importantly, conservation plans attempt to deal with biodiversity at every level of organisation from subspecies to phyla. Moreover, even if priority sites are identified for all taxa, biological processes such as ecological and evolutionary processes as well as endangered biological phenomena are not automatically included.

General Biodiversity Surrogates

Adequate representation requires that a conservation area network must include examples of all biodiversity features of a region. Biodiversity, as such, is impossible to define precisely; it refers to diversity at every level of the taxonomic, structural, and functional organisation of life. Moreover, it is presently impossible to measure all components of biodiversity in a given region. Even for well-known taxa such as birds and mammals, georeferenced information on distributions is typically incomplete. For planning purposes, features of biodiversity must be individuated and measured in some way. Consequently, partial measures, known as biodiversity surrogates, must be used. These surrogates must adequately represent biodiversity features in protocols using conservation planning tools. Although the tools themselves typically do not restrict what surrogates are used, care must be taken to ensure that adequate surrogates have been chosen in each planning context.

Surrogates that are supposed to represent total or general biodiversity are sometimes called "true" surrogates. Usually, species or other taxa are used as true surrogates. However, because general biodiversity is too diffuse a term to be precisely defined, the choice of a true surrogate set appeals at least implicitly to some convention or consensus about what constitutes the relevant features of biodiversity in a given context.

Thus, choosing a true surrogate set amounts to accepting an operational definition of biodiversity. Because complete distributional and other data on true surrogate sets are virtually impossible to obtain, estimator surrogates are used in their stead. Many types of estimator surrogates have been proposed including well-known taxonomic groups, species assemblages, spatial classifications of land and water, and environmental classes.

True surrogate set performance relative to overall biodiversity cannot be tested because there is no precise definition of overall biodiversity. However, we can test whether an estimator surrogate set adequately represents a nominated true surrogate set. Techniques proposed for such tests include predicting complementarity, use of species accumulation curves, surrogacy graphs, marginal representation plots, spatial congruence analysis, and regression analyses. These test whether the results of planning using estimator surrogate sets are the same as those that would have been obtained using true surrogate sets.

The level of support in the literature for various estimator surrogate sets has been variable. For taxonomic surrogate sets, some studies report encouraging results, but in general, there remains much skepticism about their adequacy. There has also been controversy over whether environmental features form adequate estimator surrogate sets for a variety of taxonomic true surrogate groups, though recent results are generally optimistic. The simultaneous use of both species and environmental parameters as estimator surrogates may reduce these problems.

Conservation planners must make the best use of all available environmental and biological data in order to inform decisions. Consequently, they have no option other than to use estimator surrogate sets or, at best, true surrogate sets. All planning tools we review assume the existence of adequate estimator surrogate sets; some, such as Surrogacy, provide explicit protocols for their evaluation against true surrogate sets.

TARGETED LEVEL OF REPRESENTATION

For each surrogate that must be represented in a conservation area network, planning tools typically require a targeted level of representation, for instance, the number of times a surrogate must be present or the fraction of its total area of occurrence that must be included. Additionally, in many cases, there is a targeted maximum area that can or must be set aside for conservation across the planning region. A maximum area requirement typically arises because of socially driven constraints (budgets) on set-aside land.

The existence of representation targets simplifies the design of algorithms. However, the use of targets has generated justified controversy. Targets are supposed to incorporate biological principles. At a rudimentary level, they often do; for instance, in many analyses, higher targets are set for species at risk than for those that are not. However, many common targets, e.g., 10% of the original habitat of each species, are rules of thumb adopted with no biological justification. For the total area that should be set aside for conservation in a region, a typical target is again 10%, which also has no biological basis. Pointing out the lack of biological justification, Soul´e & Sanjayan have argued that the use of targets and success in meeting them may engender unjustified confidence that adequate conservation measures are in place. Nonetheless, recent conservation plans have used comprehensive data sets and carefully formulated targets intended to reflect the relative conservation requirements of land types and species. Targets have been based on threat, rarity of species, and spatial turnover or heterogeneity of species within land types. The conservation areas designed to achieve these targets and to promote the persistence of biodiversity processes often cover very large percentages of regions (50% to 70%) and present substantial challenges for implementation.

Though traditional population viability analysis can potentially be used to set targets, there is no consensus about the appropriate framework for such analysis. Moreover, population viability analysis can only be performed on a single or very few species at a time, requires large quantities of usually inaccessible data, and yields models with a high level of structural uncertainty. There have been some attempts to formulate methods for setting targets for a large number of species. Appropriate targets for land types or environmental classes remain problematic, although some agreement can be reached on the relative sizes of targets for individual types.

Persistence of Biodiversity

To be successful, conservation plans must go beyond the mere representation of biodiversity patterns to ensure the persistence of biodiversity through accommodating ecological, evolutionary, and sociopolitical processes. These processes might include dispersal, local extinctions and recolonisations, species interactions, migration, patch dynamics, adjustment of species' distributions to climate change, and diversification of lineages. Contemporary conservation planning draws on seven sets of ideas intended to safeguard the persistence of biodiversity in a conservation area network. Planning for persistence requires, at the very least, incorporation of rules of spatial configuration that take these ideas into account.

Sociopolitical Criteria

Biodiversity conservation does not occur in a sociopolitical vacuum. Rather, designating land for conservation must compete with other social claims on land. In North America, early efforts at conservation planning often implicitly assumed that sociopolitical considerations were peripheral to biodiversity conservation for a variety of reasons, including normative claims about the equal importance of other species and humans. These discussions were subsequently criticized on both philosophical and prudential grounds. Philosophically, attempts to found ethical bases for conservation on nonanthropocentric grounds have largely been rejected, though the issue remains controversial. Prudentially, ignoring rather than addressing alternative claims on land is a recipe for political problems and likely failure of conservation plans. Most international protocols for biodiversity conservation now legitimise sociopolitical interests.

To incorporate sociopolitical criteria into conservation planning, the process must be viewed as solving a multicriteria decision problem involving criteria other than the representation and persistence of biodiversity. Techniques of multicriteria analysis are used for this purpose. Part of this broad view of conservation planning includes stakeholders-people who are affected by or who can influence decisions and those who are responsible for their implementation. Communication with a variety of stakeholders throughout the planning process has been emphasized and given prominence in some planning protocols.

Conservation Problems

The design and implementation of software planning tools require the precise specification of both the problems to be solved and the algorithms to solve them. Many of the formal problems encountered in this context have long been studied within computer science and operations research. Although conservation biologists were aware of these connections from the beginning, recent years have seen increasing involvement of operations researchers. There is scope for fruitful collaboration provided that the problems addressed remain relevant to planning in the field.

Optimisation

Most of the problems solved by conservation planning tools are formalised as constrained optimisation (minimisation or maximisation) problems. A basic problem is to look for the minimal set of sites in which all representation targets are met. What is being optimised (minimised) is the number of sites; the constraint is that all targets must be satisfied. Formalising problems as ones of constrained optimisation is useful because a wide variety of algorithms with known performance are available for their solution.

Representing the various conservation planning problems as optimisation problems involves one choice among many available options for their formalisation. For instance, if the adequate representation of biodiversity surrogates and the minimisation of the number of selected sites are viewed as goals among which trade-offs are allowed (that is, neither is a hard nor inviolable constraint), the problem is not, strictly, one of optimisation.

Conservation Area Network Selection

Many optimisation problems can be formalised as mathematical programming problems. A family of formal problems associated with the "set cover", and the "maximal cover" problems from operations research also occur in conservation area network design. The inputs for these problems are a set of sites and a list of the surrogates present in each such site. In the set cover problem, the objective is to minimise the total cost of sites such that each surrogate is represented at or above its target within the selected set. In the maximal cover problem, the objective is to maximise the number of surrogates that satisfy their targets subject to a ceiling on the total cost of selected sites. Both problems can be represented as deterministic integer programmes. Variants can be formulated to incorporate other objectives as further constraints, such as shape or minimal size of each contiguous set of selected sites. Multiple optimal solutions may exist for both the set and maximal cover problems, and standard algorithms can find all of them, allowing the possibility of comparing these solutions with respect to additional criteria.

Conservation Action Scheduling

With the sole exception of Kirkpatrick's plan for Tasmania, we are not aware of a case that implemented an entire set of selected sites. Commonly, plans are implemented partially and in stages over a planning period with the number of sites acquired at each stage constrained by a budget. Typically, planning regions are at least partly exposed to threats from expanding agriculture, mineral resource extraction, urbanisation, and other sources. So at each stage, there is a risk of losing sites within the selected set as well as unselected sites that might serve as their replacements. Under these circumstances, the goal is to take such vulnerability into account while identifying sets of sites to be acquired at each stage so that representation of biodiversity is maximised at the end of the process. This is known as the conservation action scheduling problem.

Conservation action scheduling can be represented as multistage constrained optimisation problem. The parameter being optimised (maximised) is the number of biodiversity surrogates that have met their targets at the end. The constraints are the budgets at each stage. This optimisation problem can be formalised as a stochastic dynamic programming problem with computationally intractable optimal methods of solution. It has also been formalised as a stochastic programming problem. There has been recent progress in identifying heuristic methods that give near-optimal solutions.

A heuristic approach to scheduling that has been widely advocated uses both the irreplaceability and vulnerability of sites. The rationale is that early protection of sites that are both highly vulnerable and highly irreplaceable will minimise the extent to which conservation targets are compromised before they can be secured. Several studies have addressed the interplay of conservation action and loss of areas. These studies have dealt only with biodiversity pattern; a major challenge for conservation planners is to develop strategies that combine planning for biodiversity processes with implementation constrained by land-use dynamics.

Analysis

Though the focus of existing conservation planning tools has typically been placed on the representation of biodiversity surrogates, effective planning must take sociopolitical factors into account. These can be incorporated into the planning process using multicriteria analysis. Moreover, criteria relevant to the spatial con-figuration of conservation area networks-such as size, shape, alignment, replication, connectivity, and dispersion, which are often critical to the persistence of biodiversity can also be incorporated in the same way.

Two types of protocols have been developed to incorporate multiple criteria into the selection of sites for a conservation area network. The first type consists of iterative stage protocols, in which multiple criteria are considered as each individual site or small set of sites is selected for inclusion. The second type consists of terminal stage protocols, in which multiple criteria are considered for the selection of an entire network after the formulation of a set of potential networks, all satisfying a given criterion, typically biodiversity representation. Both *iterative*- and *terminal*-stage protocols can be used simultaneously, with some criteria incorporated during site selection and others at the end.

There have been many attempts to use multicriteria analysis in the design of conservation area networks. A wide variety of methods exist, ranging from the well-developed multiattribute value and utility theories to purely heuristic procedures. Selecting an appropriate method depends on (a) whether the alternative sites or conservation networks can be ordinally or quantitatively ranked by each criterion; (b) whether the criteria for evaluating alternatives can be ranked at all and, if so, whether they can be ordinally or quantitatively ranked; (c) whether the criteria are independent of each other; and (d) whether the criteria can be compounded. Problems with ranking and compounding criteria have long been recognised in the decision theory community. Ranking criteria is often arbitrary and must be accompanied by sensitivity analyses to test whether results are unstable with even slight changes in ranking. Additionally, once criteria begin to be aggregated, there is potential for a loss of transparency; it is no longer clear what motivates a choice. Compounding has some theoretical basis within multiattribute value and utility theories. However, these theories make strong independence assumptions about each pair of criteria. Moffett & Sarkar have pointed out that existing planning tools incorporate a small fraction of the available techniques.

Data

Traditionally, conservation planning algorithms have assumed that the data available show whether a surrogate is present or absent and sometimes, if present, how abundant or extensive. These data can be obtained from systematic surveys, niche models, maps of land types, or environmental classes. A typical problem faced is that the data on species are presence only, rather than presence-absence-they show where surrogates are present but not where they have been searched for but not found. In such circumstances, sites with no data (possibly indicating false absences) may have allowed targets to be achieved more economically but do not mitigate the representation achieved within selected cells.

However, because planning is essentially a matter of comparing sites with one another, it is better to have comparable data across all candidate sites. One way to mitigate problems caused by presence-only data is to use them to model the expected geographical distribution of surrogates in the planning region. Many modelling techniques are available, and these typically report probabilities for the presence of surrogates. Similarly when the future distributions of surrogates are incorporated into a plan, we rely on models that typically make probabilistic predictions. Examples include models assessing the vulnerability of biota because of anthropogenic factors, range shifts due to climate change, and patterns of dispersion.

Probabilistic data may be converted to binary data using a threshold probability. However, this procedure is open to the objection that any threshold must be arbitrary. Two different strategies allow the use of probabilistic data directly. First, probabilities of occurrence or persistence in individual sites may be compounded to obtain the corresponding probabilities for the entire landscape. The goal is to ensure that the probability in the network is higher than some specified value, similar to a target of representation. Although this strategy is often followed, it requires the assumptions that the probabilities of different surrogates within a site are independent of each other and that the probabilities for each surrogate are independent from one site to another. Because of ecological relationships between surrogates and the spatial correlations of their distributions, both assumptions are unrealistic. Moreover, the associated computational problems are often nonlinear, although they can sometimes be linearised and made tractable.

The use of expectations also allows computational problems to be linearised. Moreover, the use of probabilistic data does not require any modification of the formalism used for binary data. The simultaneous use of probabilistic and binary data is also not problematic. That some existing tools are restricted to binary data is an unnecessary limitation.

Algorithmic Issues

In discussing algorithmic issues encountered in the design of software tools, though all criteria are important, we are concerned mainly with efficiency. The importance of computational efficiency depends on how planning tools are to be used. As decision support systems, they may be used to develop decision scenarios that require scores of alternative plans to be formulated for appraisal by decision makers. They may also be used in real-time negotiations during which stakeholders need rapid information about the implications of proceeding in alternative ways. In such scenarios, planning tools must ensure that certain minimal criteria (such as representation targets) continue to be met in each plan. Tools that take large amounts of time to analyse a data set or to produce a scenario are likely to be ineffective; in such cases, computational efficiency is critical. Conversely, in a decision scenario in which only a single or a few alternative plans are sufficient, efficiency may be less important than (spatial) economy. In recent years, an increase in the efficiency of planning tools has led to a trend toward the use of multiple scenarios. The incorporation of multiple criteria in terminal stage protocols can also involve the formulation of many alternative plans.

Complexity

Because there has been recent controversy over the efficiency and economy of different algorithms, we introduce relevant terminology from computer science. Computational complexity can either be (a) temporal complexity or the time required for a computation, with complexity being the inverse of efficiency; or (b) spatial complexity, or the amount of memory that is required for a computation. Both are relevant to the design of conservation planning tools.

With respect to temporal complexity, a computational problem belongs to the class P (for polynomial time) if the number of elementary operations (additions, subtractions, multiplications, and divisions) required to obtain an answer grows as a polynomial function of the size of the input (preferably a low-order polynomial). What is important here is that such algorithms are tractable: the time required to execute them (a polynomial function) does not grow inordinately fast as the size of the problem increases compared to an algorithm growing at an exponential rate. The class NP (nondeterministic polynomial) time consists of problems for which the number of operations required to verify a solution grows as a polynomial function of the size of the input. [The contrast here is between the time required to produce a solution (in the case of P) and the time required to verify that a solution is correct (in the case of NP).] Obviously, P is a subclass of NP. One of the most important open questions in computer science is whether P = NP.

Given these definitions, a problem is NP-complete if (*a*) it is inNP and (*b*) every other problem in NP is reducible to it; that is, any such problem can be transformed into the NP-complete problem using a P algorithm. Thus, NP-complete problems are the hardest problems in NP. Finally, an NP-hard problem is one that satisfies clause (*b*) above but not (*a*); that is, it is not necessarily in NP. Thus, NP-hard problems are at least as hard as NP-complete problems, possibly harder. The most salient aspect of NP-hard and NP-complete problems is that increasing the speed of computer processors does not significantly affect the tractability of these problems. However, this does not mean that every or even most instances of such problems cannot be solved efficiently. All it means is that there are instances for which a solution cannot be obtained in a reasonable amount of time, which is a serious constraint if the goal is to design generic software tools.

Heuristic Algorithms

For the design of conservation area networks, the set cover and maximal cover problems, which are the simplest problems to be solved, are both NP-hard. Consequently, exact or optimal algorithms guaranteed to produce the optimal (or most economical) solutions may be intractable in many instances. However, for binary data (where surrogates are either present or absent in areas), with definite targets of representation, the function to be optimised can be linearised, reducing temporal complexity. For probabilistic data, some nonlinear problems can also be represented as linear problems. Nevertheless, because of NP-hardness of even the linear problems from a computational perspective, it is important to devise efficient heuristic algorithms.

A variety of stepwise or "single pass" heuristic algorithms have been developed. For both conservation area network selection and the scheduling problem, a stepwise heuristic algorithm consists of a rule (or a series of rules) that is applied successively to select sites for inclusion in the selected set, with recalculation of dynamically changing measures such as complementarity at each iteration. Although stepwise heuristic algorithms have in practice achieved efficiency because of the simplicity of the rules incorporated in them, most of them were originally motivated by the goals of economy and transparency, which was achieved by the latter through the incorporation of biologically interpretable criteria such as complementarity, rarity, and adjacency.

The most commonly used heuristic rules that have been designed to select sites are (*a*) to maximise the complementarity of surrogates (the "complementarity rule"); and (*b*) to maximise the rarity of surrogates occurring at a site, with rarity being interpreted as the inverse of the frequency or area of occurrence of a surrogate. Extensive testing on a large variety of artificial and empirical data sets indicates that, for binary data, using both rules leads to the best result. For probabilistic data, complementarity alone performs best. The complementarity rule has been incorporated into several planning tools including C-Plan and WorldMap. Both the complementarity and rarity rules have been implemented in ResNet.

Stepwise heuristic rules are implemented hierarchically. Should one heuristic rule lead to a tie between several sites for potential inclusion, a second is used, and this process is repeated over the group of rules. For instance, if rarity leads to a tie, then an adjacency rule giving preference to a site adjacent to one already selected can be used to try to break this tie. The adjacency rule leads to the selection of larger contiguous areas. Thus, hierarchical rules allow an intuitive incorporation of multiple criteria. However, the relative importance of the rules is determined by their sequence, with frequency of rule use largely determined by the number of ties. This can lead to weightings of the rules that are not explicit.

Metaheuristic Algorithms

Metaheuristic algorithms control the use of heuristic rules by permitting their repeated use, with an independent criterion determining exit from the algorithm. Metaheuristic algorithms can also be used to incorporate multiple criteria. For instance, the selection of an initial set of sites can be followed by repeated random substitution of sites to see if better spatial configuration can be obtained without sacrificing representation targets. A time limit can be stipulated to force an eventual exit from such a metaheuristic algorithm. Metaheuristic algorithms are becoming the preferred technique for solving NP-hard problems. In the design of conservation area networks, metaheuristic algorithms potentially enable greater spatial economy than heuristic algorithms but without sacrificing computational efficiency.

There exists a wide variety of metaheuristic algorithms most of which are yet to be used in systematic conservation planning. Simulated annealing has been used extensively [with the Marxan software package and its predecessors] to incorporate spatial criteria along with representation. It is less computationally efficient than stepwise heuristic rules, but the two can be used together to achieve spatial coherence of network design without excessive loss of computational efficiency. C-Plan (with heuristic rules) has recently been adapted to work interactively with Marxan. Tabu search has also been used and shows promise.

Metaheuristics can provide a solution to the limitations of sequential applications of rules in stepwise heuristics. For instance, simulated annealing can be used with an objective function, incorporating achievement of targets, minimisation of costs, and a function to achieve spatial compactness. These criteria are then implemented simultaneously, not hierarchically, and their relative influence on the solutions can be adjusted using weights. However, such weights may be criticized as being arbitrary, and sensitivity analysis should be used to test the robustness of results.

Optimal Algorithms

Optimal algorithms necessarily perform better than heuristic and metaheuristic algorithms with respect to economy, but because the problems are NP-hard, these algorithms suffer from poor computational efficiency; that is, they may take inordinate amounts of time to resolve realistically sized data sets. Consequently, the gain in economy may not offset the cost in efficiency and the lack of transparency in most practical contexts. Numerous computational studies have analysed the efficiency and economy of stepwise heuristics compared to optimal algorithms. Heuristic algorithms achieve computational efficiency and transparency at a potential loss of economy, although this loss has not been rigorously quantified. Extensive numerical tests in the 1990s underscored the conclusion that, compared to heuristic rules, optimal algorithms only achieve a minor increase in economy with a significant loss of efficiency.

Rodrigues and coworkers questioned these results and claimed that new software solvers produced optimal solutions with as much computational efficiency as heuristic algorithms. Sarkar et al. examined these claims systematically. Using the industry-standard CPLEX and other integer programming solvers, they found more efficient optimal solutions than Rodrigues & Gaston. However, Sarkar et al. found that heuristic algorithms uniformly outperformed optimal solvers with respect to efficiency with only a marginal loss of economy for realistically sized data sets, especially with probabilistic data. In an extreme case, with a data set from Ecuador, the CPLEX and XPRESS optimal solvers found solutions with 65 and 56 sites in 16,018 and 532 seconds, respectively, whereas heuristic rules found solutions with 50-62 sites in 17-23 seconds. Rodrigues & Gaston's negative results for heuristic rules were apparently due to poor implementation of heuristic algorithms. Importantly, they also reflected the simplistic nature of the problems they addressed. Most involved binary data and a target of one representation of each surrogate. Problems of this sort have little practical relevance.

For the scheduling problem, optimal algorithms have also proved computationally intractable because of spatial complexity. The standard method for stochastic dynamic programming is to use backward recursion. However, the number of possible states of the landscape (which sites are selected, which are available, and which have been lost) is so large that the computation becomes impossible for data sets with more than about 25 sites. Recent attempts to solve the scheduling problem have, therefore, relied on heuristic methods. Some of these have been tested using realistic simulations of parallel, incremental conservation and loss of native vegetation.

Priorities for Future Research

To date, conservation planning tools have only implemented relatively simplistic decision support protocols compared to the large array of methods available. Most conservation decisions are typically made by groups of stakeholders. Nonetheless, existing planning tools, with very few exceptions, only support individual decision making. Although a variety of methods exist for group decision support, very little work has been done to examine the suitability of these methods for conservation planning. Performing such an analysis and the subsequent design of appropriate tools remain important future tasks.

Similarly, planning is an iterative process with the results of initial planning exercises being used not only to determine what further data and analyses are necessary, but also to refine and devise new policy alternatives. Several software tools are designed to be used iteratively, but they do not fully implement the iterative process in the sense that they do not provide explicit protocols for the formulation of new policy alternatives. Rather, that task is left to the user. The next generation of planning tools should provide additional support at this stage.

It is also widely recognised that conservation decision making occurs under conditions of risk and uncertainty. Some forms of future risk and uncertainty are beginning to be addressed in conservation planning tools. There has also been extensive work on these topics in the context of the management of individual species, both from classical and Bayesian viewpoints. However, much more remains to be done even in these contexts. Other forms of uncertainty-for instance, dynamic changes in the available sites or changes in criteria-have had limited consideration in the design of conservation planning tools.

Finally,conservation planning must occur simultaneously at multiple spatial scales. Moreover, the biodiversity features that must be conserved include groups from all taxonomic levels with varying amounts of reliable information available at each level. Thus, conservation planning tools of the future must incorporate systematic methods for dealing with spatial and taxonomic heterogeneity of scale in the data available. A related problem is that the spatial and taxonomic scales at which analyses may be reliably performed, as constrained by the available data, may not be the scale of implementation. Few studies have compared the distribution of priority areas derived from coarse-scale (regional) and fine-scale (local) data, but these have demonstrated significant differences. At many spatial scales, decisions are likely to be under the purview of groups rather than individuals, and the size, composition, and organisation of groups are likely to vary with geographical scale. Different types of uncertainty and risk pertain to different scales and percolate between scales in ways that are difficult to quantify.

Issues connected with the choice of targets and surrogates have been recently reviewed in discussions about biodiversity data for conservation planning. We mention two important unresolved issues.

— Partial protection-planning tools should take into account contributions from partially protected off-reserve areas. This requires explicit information on the likelihood of persistence and contribution to targets for each surrogate under different management regimes. Although conservation planning tools can use such data (in binary or probabilistic form), these remain difficult to obtain. Problems of predicting surrogate distributions and persistence likelihoods relative to a variety of management regimes remain unsolved.

— Beyond species to other levels of variation-phylogenetic diversity (PD) has long been incorporated in planning tools, but it has not yet had much impact on conservation planning. Applications face limitations of available data on phylogenetic pattern. In the near future, new approaches may vastly increase available information for many taxa and many sites. For example, it has recently (and controversially) been suggested that a mitochondrial gene (cytochrome c oxidase subunit I, COI) can potentially distinguish related species across a broad range of taxa, and rapid COI assessment may enable a general "barcoding" approach for assigning unidentified individuals to species. Application of conservation planning tools could benefit from such information. Other promising prospects include the use of PD for rapid conservation assessment by applying PD measures to molecular (for instance, COI-based) phylogenetic patterns.

Biological justification of representation targets is often lacking, although they do offer bench marks for comparison of representation achievement. Targets can be avoided to some degree in many cases. For instance, marginal representation methods, such as those used for surrogacy analysis by Sarkar et al. do not use targets. A range of little-explored approaches exist that use continuous surrogates (as, for instance, obtained using ordination) and do not use targets. Applications include those that explore shifts in frontier curves under different scenarios.

One approach was explored in conservation planning for Papua New Guinea. The amount of continuous variation that could be represented in 10% of the total area with no constraints was recorded. This amount was then matched in the set of priority areas identified in the presence of fixed existing protected areas, costs, and other constraints. Such an approach required an arbitrary target only for use as a bench mark.

There has been only one case in which an entire set of selected sites has been fully implemented. An alternative to the selected set approach is to develop a continuously scaled measure of conservation value that can be constantly updated as the conservation landscape changes. All sites are prioritised by their current value, and a site can still be acquired from those that are currently available to maximise biodiversity representation within the available budget. Having a continuous conservation value for every site also has an educational benefit by showing that most sites have value, whereas the selection of sets may unintentionally convey the message that sites not in the set have no value.

Finally, the dynamic realities of planning may put targets, discrete attributes, and sets into question to such an extent that some have questioned the use of computer algorithms. Policy-based algorithms may mimic the iterative approaches of multicriteria analysis, as in ongoing selection of landowners for conservation payments in Western Australia. Tools for strategies such as these remain to be developed in the future.

Biodiversity Processes

For much of its development, systematic conservation planning has been dominated by considerations of biodiversity patterns or features, such as species locality records and vegetation types, that can be mapped, treated as static, and sampled in conservation areas. But an important role of conservation areas is the maintenance of ecological and evolutionary processes.

Although quantitative targets for pattern-based features can go some way to promoting the persistence of processes, the conservation area networks that result are unlikely to support biodiversity processes that require large areas or particular spatial configurations. Four types of approaches have been envisioned to promote the persistence of processes. First, moveable priority areas can track movements of biota of interest or resources necessary for their survival.

In terrestrial environments, such areas are generally easiest to consider and implement if they cover publicly owned small areas because of potential constraints imposed by protective management of extractive uses in privately owned land. Moveable areas can be valuable for protecting features such as bird breeding colonies, bat roosts, or populations of disturbance-dependent rare plants that periodically shift within a region. They have also been proposed to protect successional processes after disturbances. A special case is the temporary but repeated application of management restrictions or hunting closures to specific areas to protect species when they are particularly vulnerable. In the marine environment, where use rights are generally less restrictive, there is more scope for shifting priority areas, even extensive ones, to track features of interest or to apply temporary protection.

Second, biophysical templates associated with specific processes may be targeted for protection. Some recent published work comes from the Cape Floristic Region of South Africa. One example is the protection of river gorges important for movement of plants and animals between regions and as refugia. A less obvious example concerns interfaces between acid and alkaline soils believed to be associated with historical and ongoing diversification of some plant taxa. Most regions have areas that are important for the persistence of species of interest. The reasons might include habitat quality and associated high densities and/or reproductive output or areas that include drought refugia, defined by topography and drainage characteristics, and stopover sites for migratory birds. Templates can also be defined as combinations of features that provide complementary habitats used at different times or for different functions, e.g., feeding, breeding, or roosting. Information on templates that represent dispersal barriers, such as wide rivers or mountain ranges, can be used to subdivide regions to reflect the influence of ecological and evolutionary history on biotic composition. Lists of templates and methods for identifying them are likely to vary widely between regions owing to the idiosyncrasies of species' life histories, biogeographic history, climate, and physical environments.

Third, qualitative and quantitative design criteria may be used to indicate preferences of planners when choices are available. Design criteria include size, shape, connectivity, replication, spacing, and width of buffers in conservation area networks, which all influence the persistence of species and the maintenance of processes. When applied qualitatively, they are expressed as preferences (e.g., bigger is better). Qualitative criteria have influenced decisions about the design of conservation areas for decades and have guided assessments of existing reserves. Preferences for higher compactness and connectivity have been implemented with rule sequences in heuristic algorithms or through objective functions. A key limitation of these approaches is that, because they are qualitative, there is no explicit link between the expressed preferences of planners and the configuration requirements of individual processes. Quantitative design criteria interpret the spatial requirements of particular processes as explicit targets. These criteria include those listed above, but the requirements are stated quantitatively. They require planners to interpret knowledge of specific processes as quantitative targets for size, connectivity, and other considerations sufficient to promote the persistence of those processes. This is difficult and constitutes an area of innovative research in conservation planning.

There is some overlap between the methods grouped under these categories. Most existing planning tools allow at least partial incorporation of qualitative design criteria, and some can use quantitative design criteria. But none systematically implements all features associated with the persistence of processes.

Relative Vulnerability

The purpose of conservation areas is to mitigate at least some of the processes that threaten biodiversity. Incorporating information on threatening processes and the relative vulnerability of areas and features to these processes into planning tools is therefore crucial for effective conservation planning. Pressey et al. defined vulnerability as the likelihood or imminence of biodiversity loss caused by current or impending threatening processes. Wilson et al. referred to this as "exposure," one of three dimensions of vulnerability in their broader definition. Their other two dimensions are the intensity of a threatening process and its impact, reflecting the response of species or other biodiversity features. Areas of particular concern for conservation planners have high exposure to highly intense threatening processes. Features of concern are those occurring in such areas and experiencing strongly negative impacts. Polasky et al. deal with all three dimensions of vulnerability.

Spatially explicit predictions about vulnerability most commonly deal with exposure because intensity and impact are more difficult to map. Exposure is related to environmental factors, e.g., topography and fertility, and to spatial factors, e.g., infrastructure or source areas of invasive species. Conservation areas may also attract development because of the added amenity value of being adjacent to perpetual open space, which also increases the land value. Many methods have been used to predict exposure, but there have been few comparisons of their results and little or no validation of predictions against actual events.

Opportunities for Benefiting Biodiversity Conservation

Companies operating in areas of high biodiversity value are increasingly expected to go beyond simply mitigating the potential adverse effects of their operations and make some sort of positive contribution to biodiversity conservation. By working closely with government officials and other local stakeholders and carefully evaluating the local economic, environmental and social situation in a project area, companies can develop effective programmes and strategies for benefiting biodiversity conservation in the areas and countries in which they work.

Companies can make investments in biodiversity conservation at both a project level and a company level. At the project level, such activities are likely to be strongly driven by the results of a project Environmental and Social Impact Assessment (ESIA) and any identified value associated with actions that go beyond mitigation to benefit valuable and threatened ecological resources. At the company level, opportunities to benefit biodiversity conservation can be a key part of an overall company environmental and social responsibility strategy that recognises the strong role of biodiversity conservation in sustainable development and the business value of a positive public reputation on biodiversity issues.

By proactively capitalising on opportunities to improve the state of conservation in ecosystems with high biodiversity value, companies can ideally leave an area's biodiversity, or the local capacity to conserve it, in better condition than before oil or gas activity began. Such activities are distinct from offsets, which are designed to reduce and compensate for the negative impacts of a project. While activities designed to benefit biodiversity may be similar to those intended to meet no net loss goals, they often encompass a broader geographic area or timeframe. Investment in an opportunity to benefit biodiversity assumes that shortterm loss of or damage to biodiversity as a result of oil and gas production can lead to a long-term net benefit to both the economy and the environment through the reinvestment of economic rents from the oil or gas activities, both from government revenue and company donations, into conservation beyond the life of a project.

Technologies and techniques for mitigating many of the primary impacts of oil or gas development are well known throughout the energy industry. Criteria for selecting the best opportunities for benefiting biodiversity conservation, however, are neither the norm nor subject to the same standard decision-making procedures as prevention and mitigation of impacts.

Local, Regional and National Conservation Strategies

Companies often find themselves caught among local communities, national governments and other parties such as environmental NGOs that have different, often competing and sometimes contradictory visions for how land and natural resources in and around a proposed project area should be used. As such, determination of the most promising opportunities for benefiting biodiversity in a given area or country must take into account existing or planned local, regional and national conservation strategies and priorities and the interests of relevant stakeholders.

Countries that have signed and ratified the 1992 U.N. Convention on Biological Diversity (CBD) are required to develop a National Biodiversity Strategy and Action Plan (NBSAP) to identify how they will achieve the biodiversity goals set forth in the CBD. Review of country-specific NBSAPs can help a company identify both priority areas for host countries and what opportunities exist to help achieve national biodiversity goals. Moreover, NBSAPs can identify important national stakeholders whom companies could work with in designing and implementing specific biodiversity conservation projects.

In some cases, a regional land-use planning exercise may have taken place or be underway to identify stakeholders and their interests and priorities and determine ways to factor those interests into regional long-term development plans. Information from such exercises can be invaluable for determining the best use of resources designated for conservation benefits. If a company can participate in or even help to initiate such a process, the experience may help to defuse potential tensions, increase trust and credibility among stakeholders and improve the ability of managers to make sound investment decisions. There are risks to such a process, however, including a long lead-time and the potential for lack of interest or even opposition from governments or other stakeholders. Such factors should be included in a company's risk management decisions.

Risks and Benefits

There are a number of different options for investments that can benefit biodiversity conservation in a given area. Each option will carry unique risks and benefits to each company. When deciding which type of option to pursue, project managers should assess the potential strategic value to the company of a particular choice, in terms of local stakeholder relations, potential effects on project performance and execution, and broader public reputation. These benefits then need to be weighed against the possible risks of a specific choice, such as the potential for conflict, difficulties in measuring effectiveness, local capacity to manage funds or project activities, or likelihood of efficient use of resources. Furthermore, in order to be effective, an investment in biodiversity conservation should address a real conservation need and will typically need to be long-term. Because this may increase a project's costs and exposure to risks, choices about opportunities should be factored in up front in analyses on both financial and reputational risks and benefits. It can be very costly to reputation to stop a project after it has been started, if it cannot be sustained.

Local Partners

The most appropriate long-term benefits for biodiversity can be best identified by working with local partners who know the status of local or national ecosystems and can point to what measures would effectively promote conservation. Potential partners include, but are not limited to, government agencies, local communities, conservation NGOs and other private sector actors, such as timber concessionaires. Local, national or international NGOs can serve as partners in bringing various stakeholders together in a consultative process. Many NGOs have substantial experience working with other local stakeholders, such as communities, and may have extensive knowledge of an area's biodiversity and what actions are needed to conserve it.

Including relevant stakeholders in the decision-making process is critical for long-term success of conservation actions. For example, a government protected areas agency may want support to increase its capacity to manage a national park, but a nearby community that is making incursions into the park to support their livelihood may want economic development assistance. Instead of viewing such a situation as "zerosum," companies can work with stakeholders to find truly sustainable solutions that increase government capacity to manage the protected area and address the development concerns of the community.

Biodiversity Richness

The status of species in an area, including endemic and threatened or endangered species, is a central issue in determining what actions may be best for benefiting biodiversity conservation. Baseline studies and analysis of secondary information to determine the status of an area's biodiversity are crucial in this regard and should be included in a company's ESIA process. At a project level, if the area in or adjacent to a project site is determined to contain valuable biodiversity, local measures, such as protection and capacity-building with local stakeholders, may be most appropriate. If a project site does not have particularly high biodiversity value, but there are other areas of the country or high-profile species that are important and threatened, a company could still consider making contributions to protect those ecosystems or species through support of local partners.

Both efforts contribute to sound biodiversity action and can improve a company's public image within the country and add value to its reputation.

Degree of Threat to Biodiversity in an Area

The overall degree of threat to biodiversity in an area should also be considered when assessing opportunities to benefit conservation. Threats to biodiversity can originate from the primary impacts of a project, but, often, secondary impacts (the unintended consequences of a project such as in-migration or illegal logging) are the most destructive long-term impacts to biodiversity from project activity. In addition, an area's biodiversity might already be under severe threat from other economic activities or social trends. In assessing the level of threat to a particular area, the following factors should be considered:

— Potential primary impacts of a project site, such as land-clearing, waste discharges, etc.

— The economic importance of the site and the region.

— Present and projected economic activities, both legal and illegal (for example poaching or illegal logging).

— Population pressures.

— The opportunity cost of protecting areas identified as having high biodiversity value.

— The land tenure of local communities.

Protected Areas System

The process for deciding what are the best opportunities for a company to benefit biodiversity should consider the legal framework and current status of a country's protected areas system. Some countries, such as Costa Rica or South Africa, have fairly well-developed protected areas systems whose needs would differ from those of a country with a poorly developed or nonexistent system. In countries where the protected areas system is poorly developed, and the protected areas are simply "paper parks" with little real management capacity, consideration can be given to supporting increased capacity through the development of better park infrastructure and more training and resources to attract and maintain qualified staff.

In those relatively few countries with no legal framework for protected areas, other opportunities can be explored, such as working with communities living in or around ecosystems identified as having high biodiversity value to promote conservation, or supporting training programmes in improved resource management. Engaging stakeholders, such as government agencies, communities and NGOs, with interest in a particular area is the best way to determine what types of support would be most effective.

Analysis of Technical and Management Capacity

Analysis of technical and management capacity at the local and national level is important in assessing opportunities for benefiting biodiversity. A region or country may have many laws and government agencies that address the issue of biodiversity protection, but little or no technical capacity, such as biologists or resource management specialists, to enable such laws and agencies to be effective. Conversely, technical capacity may exist to manage biodiversity issues, but funds might be inadequate to attract and keep the type of personnel needed. In the former, technical capacity building may be the better opportunity for a company to benefit biodiversity, whereas in the latter, financial support for better utilisation of existing capacity, in the form of salaries or funds for equipment or infrastructure, may be a better option. To avoid creating conflicts of interest, funds earmarked for public employee salaries can be disbursed through endowments or trust funds managed by entities other than the company making the contribution. While capacity-building measures, such as technical training, may not deliver immediate, easily measurable results they can still be significant for long-term conservation. Any local or national system designed to address biodiversity issues will ultimately need qualified staff to collect baseline data, monitor changes over time, and provide analysis and recommendations.

In some countries, the issue of biodiversity conservation may be politically contentious. At the local level, communities may view biodiversity conservation measures as efforts to deprive them of access to traditional or ancestral lands and the resources they provide. Failure to take into account community considerations could lead to conflict and ultimately ineffective conservation efforts. National governments opposed to biodiversity conservation measures or more interested in development issues may be more receptive to support for sustainable development programmes or technical capacity-building measures. In these situations, companies can work with stakeholders to devise economic development options that are more compatible with biodiversity conservation, such as better land-use planning.

Conversely, a company may operate in a country where conservation is a priority for governments and communities. In such cases, leveraging stakeholder support can be valuable for optimising opportunities. Support from indigenous communities for the legal demarcation and protection of their traditional lands in countries such as Brazil has led to the protection of significant areas from more destructive land-use practices, such as logging and colonisation.

Examples

There are many ways that a company can take action to benefit biodiversity near a project or at the regional or national level, based on the most outstanding needs and problems related to biodiversity conservation in the area. The following are types of opportunities that a company might take advantage of, including examples of cases where companies took proactive measures to benefit biodiversity as a result of implementation of their oil and gas development projects. The opportunities presented here are by no means an exhaustive list, as the local and national conditions and criteria set forth above will ultimately determine which opportunities are best for a company and biodiversity conservation.

Protected Areas

Protected areas are critical tools for effective long-term biodiversity conservation. While protected areas cannot on their own ensure conservation, in combination with improved resource management and sustainable economic development activities, they can form the centerpiece of a country's biodiversity conservation strategy. Companies interested in benefiting biodiversity conservation by supporting protected areas have a number of options to consider including:

Many national parks and other protected areas, particularly in poorer countries, lack sufficient resources to be managed effectively. Establishing a trust fund or making contributions to an existing fund for park management can be the most effective way to guarantee a long-term revenue source for a protected area. In countries where the legal framework does not allow for a trust fund mechanism, direct annual payments or lump sums can be used, as long as there are adequate controls for accounting and disbursement of funds. Support for capacity-building in fund management should be included to ensure resources are effectively managed over the long term. Strategic in-kind contributions, such as patrol vehicles and infrastructure for better management (observation posts, patrol routes, etc.) can also be effective in increasing protected area management capacity. Stakeholder consultation and partnerships will be critical in determining the resources needed for effective protected area management and how those resources are used.

In Bolivia, Shell International worked with the Foundation for Friends of Nature, the Foundation for Friends of the Noel Kempff Natural History Museum, the Wildlife Conservation Society, the Missouri Botanical Garden and Enron to establish The Chiquitano Forest Conservation Foundation (FCBC in Spanish) in September 1999. A private non-profit organisation with independent administration, the FCBC was established to support biodiversity conservation efforts in the eastern part of the Chiquitania region of Bolivia and prevent environmental impacts from regional development and large infrastructure projects in the region. Private voluntary contributions form the financial basis of the FCBC and are distributed to the region's stakeholders at the rate of US$1 million per year. Over the course of 15 years, another US$1 million per year will be placed in a trustee fund that guarantees the sustainability of the FCBC after active contributions have ended.

Much of the world's most valuable biodiversity is not under any type of formal protection. If unprotected ecosystems with high biodiversity value are identified near a project site, a company can consider working with stakeholders to add new areas to the host country's protected areas system. Moreover, to enable a new area to be effectively managed, a company should work with stakeholders to ensure that adequate resources are available for effective long-term management.

In 1996, Mobil acquired an exploratory oil concession in the Tambopata Candamo Reserve Zone (TCRZ) in Peru, a 1.5 million hectare rain forest ecosystem that lies between the Bolivian border and Peru's famous Manu National Park. The Tambopata region holds some of the most pristine and unspoiled ecosystems of Amazonia and the highest single site species diversity records for woody plants, birds, butterflies, mammals and dragonflies. In 1998, Mobil decided to end exploration activities and leave the area, but before exiting, the company worked with the Peruvian Government and Conservation International to add the TCRZ to the existing Bahuaja-Sonene National Park, doubling the park's size to 1.1 million hectares. Returning the concession also led to the creation of the Tambopata National Reserve and adjacent buffer zones.

In June 1999, BP Petronas Acetyls, a joint venture between BP and Petronas, partnered with the Malaysian Department of Fisheries and the World Wide Fund for Nature Malaysia to create the Ma'Daerah Turtle Sanctuary Center in the state of Terengganu, Malaysia. BP has three petrochemical plants in Terengganu and there are significant oil and gas reserves off the east coast of the state. Terengganu is home to about 70 percent of Malaysia's turtles and the sanctuary is an important nesting habitat for three species of marine turtles and the painted terrapin. It is the first turtle sanctuary to be funded by the private sector and the second largest sanctuary in Malaysia. The center's main goals are to improve management of the habitat, improve community awareness, support scientific research and monitoring, and undertake controlled ecotourism projects.

Production activities often require only relatively small tracts of land within a larger concession or project area. If a concession contains ecosystems determined to have high biodiversity value, the portion not needed for operations can be managed as a formal or de facto protected area. The company can assume direct management responsibilities, or have other entities, such as government agencies, manage the area. Where communities and other civil society entities, such as NGOs, are present, they should be included in the design and implementation of a management plan. In some cases, biodiversity inside the boundaries of a company's concession may be healthier than that outside the boundaries, because of the company's ability to prevent human incursion and activities that lead to environmental degradation.

Caltex Pacific Indonesia has been extracting oil from the Zamrud Field, in eastern Sumatra, Indonesia, since 1982. The concession is in an area of lowland coastal virgin rainforest and contains two pristine lakes. When the company discovered oil in the area in 1975, it decided to manage the concession as a conservation area, a decision supported by the Government of Indonesia, which officially designated the area as a national conservation area. This decision had significant consequences for project design and implementation, leading to the development of minimal impact seismic surveying, zero discharge policies and an extensive monitoring and assessment programme.

Charismatic and Endangered Species

Efforts to protect flagship species, such as tigers or orangutans, can be the key to protecting an entire ecosystem. Companies can identify charismatic and endangered species located near a project site or in another part of the country or world, and contribute to efforts to protect it. It is important that such efforts focus on in situ conservation, protecting not only individuals of the species, but the critical habitats upon which they rely for survival. This approach has the dual conservation benefit of not only protecting a specific species, but also all the other biodiversity found in its habitat. Moreover, companies can often use species-specific programmes to build their public reputations, such as ExxonMobil has done with its "Save the Tiger Fund," an effort to protect Asia's endangered tiger populations.

BP in Spain is supporting projects to develop and maintain a permanent refuge for the Iberian lynx, the world's most endangered wild cat species. The lynx, which once roamed throughout the Iberian Peninsula, is now extinct in Portugal and only 150 to 200 individuals remain in southern Spain. The principal threats to the cat are disappearance of its main prey, the European rabbit, and fragmentation and destruction of its habitat. BP is working with the Global Nature Fund, with advice from the World Wide Fund for Nature, to support the needs of the lynx and restore lynx habitat in southwestern Spain through financial contributions and public education and awareness campaigns that include sales promotions and point-of-purchase materials such as posters and displays. The first habitat project of the campaign is the adaptation of La Finca del Gato, an area near Doñana National Park, into a protected lynx reserve. In addition, a Global Nature Fund project is working to re-establish the depleted European rabbit population, create new water sources for both lynx and rabbit, and restore the habitat essential for continued lynx survival.

Scientific Research and Analysis

Many countries lack the national and local expertise needed to effectively gather data for assessing, monitoring and protecting biodiversity. Companies can make significant contributions to a country's scientific capacity to manage biodiversity by supporting research and training in biodiversity-related fields. Inclusion of partners, such as NGOs and universities, in the ESIA process, baseline data collection and monitoring can often add significantly to local research and analytical capacity. Support for training and capacity-building have the additional advantage of often being viewed as "apolitical," and could be an attractive option in countries where the issue of biodiversity conservation may be politically sensitive.

Shell Prospecting and Development (Peru) contributed both money and resources to a comprehensive baseline biological analysis of a previously unexplored area of primary tropical forest at its Camisea natural gas project in the Peruvian Amazon. Before ending its involvement with the project in 1998, Shell contracted the Smithsonian Institution's Monitoring and Assessment of Biodiversity Programme to do a biodiversity assessment of the region, in order to predict potential impacts of development and create a baseline of information to use in long-term monitoring. The project involved more than 100 Peruvian and international scientists studying species and ecological functions, including aquatic systems, vegetation, invertebrates, amphibians and reptiles, birds and mammals. The studies revealed a tremendous diversity of species, and a number of plants and animals that were previously unrecorded.

ChevronTexaco provided accommodation and an office to whale researchers on their Lombo East platform, off the coast of Angola. A team from the Whale Unit of the Mammal Research Institute at the University of Pretoria, South Africa, used the platform in 1998 as a base to study the migration and activities of humpback whales.

Statoil, in cooperation with the Norwegian Institute of Marine Research (IMR), has supported the study of reef-building coldwater corals off the coast of Norway. The partnership has led to the identification and mapping of a number of coral reefs in the cold waters off the Norwegian coast. Documentation of damage to several reefs from trawling led to the protection of one the largest coral reefs in Norway (the Sula reef) against trawling in 1998. An inshore coral reef in the Trondheim fjord was preliminarily protected as the first Norwegian marine nature reserve in 2000. Statoil surveyed the area in cooperation with IMR to identify a possible pipeline corridor, leading to the modification of proposed pipeline routes.

Shell Gabon is working with the Smithsonian Institution's Monitoring and Assessment of Biodiversity Programme to improve knowledge and management of biodiversity within the Gamba Complex in Gabon. The partnership includes biodiversity research, assessment and monitoring; promotion of links among researchers, conservation scientists and resource developers in Gabon; technical training to increase in-country capacity for continued biodiversity assessments; broad dissemination of scientific information generated from the biodiversity assessments; and the development of partnerships for conservation and sustainable development among local stakeholders, scientists and industry.

Environmental Awareness

Lack of public awareness of biodiversity and its importance can be a significant impediment to successful long-term conservation. In areas where knowledge of biodiversity conservation issues is poor, support for public education and awareness campaigns can produce understanding and support among populations that may have previously been hostile or indifferent to the issue. Environmental education and awareness programmes also have the advantage of being scalable to the level needed, from specific communities living near areas of high biodiversity value, to nationwide campaigns to promote greater support for conservation.

In April 2001, BP sponsored an Environmental Awareness Week to raise awareness among young people in Azerbaijan. Implemented in partnership with several local environmental organisations, the week aimed to highlight current international environmental problems and encouraged everyone, regardless of age, to take more responsibility for environmental challenges. Classes covered the history and impact of oil production in

Azerbaijan, current local environmental conditions and the latest technologies used for the protection of the environment. A number of leading Azerbaijani environmental scientists from the Caspian Environmental Laboratory, which is now operated by BP, taught classes and took children on field excursions.

Shell Bangladesh Exploration and Development BV has supported the Center for Natural Resource Studies, a local NGO, in efforts to conserve the sea turtle population in Bangladesh through participatory activities involving local villagers, schools and relevant governmental agencies and non-governmental organisations.

Sharing of Information

Private sector companies, particularly multinational corporations, hold enormous amounts of biodiversity data within their archives - data that could be valuable for the wider scientific and biodiversity research community. Much of this data is generated during the preparation of ESIAs. Biodiversity studies are frequently undertaken to support ESIAs or other environmental activities and biodiversity monitoring often continues throughout the lifecycle of a project. While some of this information may be confidential, much of it could be shared with other parties. Currently, there are no widely used mechanisms to make these biodiversity data accessible, even at the most basic level. Making this information available and accessible locally, nationally or even internationally can greatly contribute to existing and future efforts to understand and conserve biodiversity.

Capacity-building in Government Agencies

In some cases, government agencies may lack the governance capacity to effectively oversee and implement national biodiversity action plans. Companies can support skill sharing, technology transfer, training and education programmes to increase the ability of government representatives to design and implement environmental policies and legislation that support national conservation priorities.

Fauna & Flora International (FFI) has been working with the governments of several countries in the Caspian region, including Azerbaijan, Iran, Kazakhstan and Russia, to develop their National Biodiversity Strategies and Action Plans (NBSAPs), as required by the Convention on Biological Diversity. BP and Shell have provided financial support to this process. In addition, the companies are working with the

Caspian Environmental Programme in Baku, Azerbaijan, and FFI to support the establishment of a Caspian Biodiversity Strategic and Policy Unit to plan for the sustainable development of the shared resources of the Caspian. This body will help develop a Biodiversity Protocol to the Caspian Sea Convention and ensure that activities at the Caspian regional level are in compliance with the international obligations such as NBSAPs and that their identified priorities are compatible with Caspian-wide priorities.

Conservation Easements

Many governments in the developing world have an abundance of forests, but few resources to meet the needs of their populations. As a result, large swaths of forest, many which are have high biodiversity value, have been concessioned off to logging or agricultural companies at very low prices, often only a few dollars per hectare. Here, a company can purchase land or the rights to a forest concession and provide funding to compensate the government. The purchased land or concession can then be managed to conserve biodiversity in cooperation with local communities, conservation organisations and other relevant stakeholders. This option would be particularly effective if land or concession rights adjacent to or in existing protected or sustainably managed areas were purchased, as they would create additional areas to protect biodiversity.

In 2001, Shell pledged US$225,000 to The Nature Conservancy to support the hiring of a new Land Conservation Representative to identify and secure properties and easements on ecologically important lands on the Alberta Rocky Mountain Front in Canada. In 2002, after the first full year of work, the new staff person had completed land conservation deals worth more than US$2 million.

Integrated Conservation and Development Activities

The presence of an industrial project in a poor, inaccessible area often creates expectations in communities that the company will provide the resources needed to lift them out of poverty. Companies can make significant contributions to the well-being of nearby communities, but some contributions, such as roads, can have both primary and secondary impacts on biodiversity. Company support for integrating conservation and development activities at the appropriate scale (i.e. regional land-use planning) can help communities promote development without compromising the integrity of ecosystems. Designing and implementing economic alternatives such as agroforestry, ecotourism and improved resource management can promote sustainable community economic development without threatening local biodiversity. Indeed, switching from practices such as slash-and-burn agriculture to agroforesty methods may actually improve the status of biodiversity in a particular area by creating buffer zones or increasing connectivity between ecosystems. The role that regional land-use planning plays is crucial, as it allows for a broader analysis and development of options to achieve both development and conservation goals in the area in question.

At its Kutubu joint venture oil development in the highlands of Papua New Guinea, ChevronTexaco is partnering with the World Wide Fund for Nature (WWF), national and provincial government and local landowners to implement the Kikori Integrated Conservation and Development Plan. The initiative, which began in 1994, includes a major biodiversity study of the region by both local and international scientists, development of pilot ecoforestry and ecotourism projects to lessen pressure on the standing forest, raising community awareness about the negative impacts of industrial-scale logging, and training for local government officials and community members in conservation skills and management. In 2001, the company joined with WWF to create the Community Development Initiative Foundation to support sustainable social and economic development in surrounding rural communities while protecting biodiversity. The foundation focuses on health care, education, agriculture, development planning, sustainable use of natural resource and capacity-building.

Trade-offs in Biodiversity Conservation Planning

Biodiversity assessment is a form of risk analysis in that it involves decision-making based on quantifying the unknown or partly known. There is consequently an element of openendedness to processes such as selection of protected areas; any additional area protected will always increase estimated overall biodiversity protection to some degree. Trade-offs are required, given the realities of limited resources and the competing demands of society. Consequently, not only biodiversity information but also information about other needs of society must be considered. While such trade-offs have been considered at the level of species priorities. Trade-offs generally may rely on cost-benefit analysis or, alternatively, more general multi-criteria analyses that escape the need for economic/dollar valuations of biodiversity.

Trade-offs that are concerned with the biodiversity of different areas pose special problems because the biodiversity value of an area changes with the changing status of other areas. The biodiversity value of a given area is typically expressed in terms of the components of biodiversity that it has that are additional to the components protected elsewhere. That "marginal gain", to use the economists' term, is called the *complementarity* value of the area.

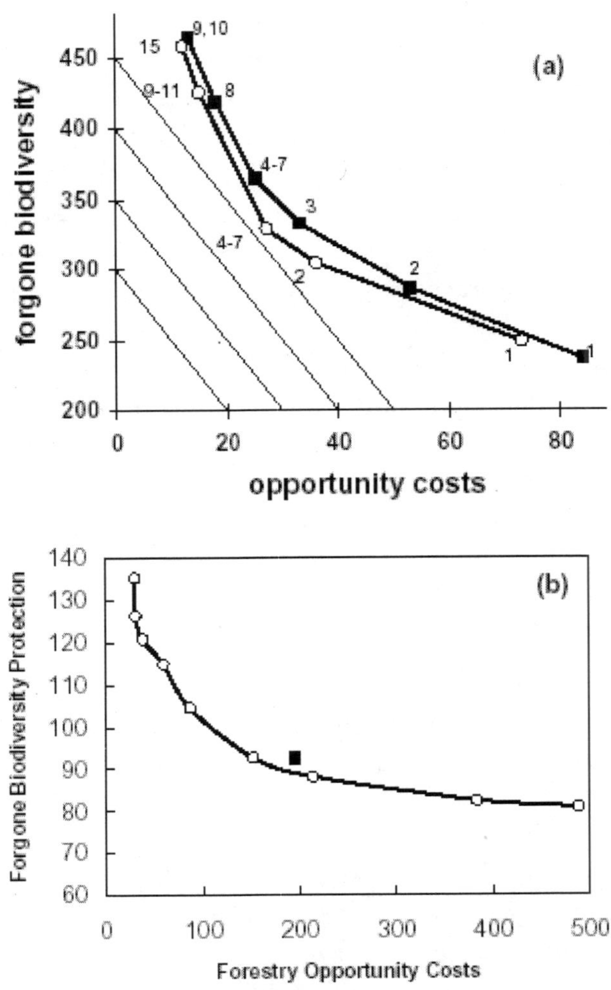

Figure 1. (a) A figure re-drawn from Faith showing two trade-offs curves in a trade-offs space. (b) A re-drawing of the trade-offs space from a study in NSW, Australia.

Regional biodiversity planning can take advantage of complementarity in a way that promotes "regional sustainability" – a balance between competing needs of society. Areas selected for biodiversity protection should have a high enough complementarity value to compensate for any corresponding "opportunity cost" of conservation (the cost to society implied by forgoing some other land use in favour of biodiversity protection).

Effective comparison of complementarity and costs depends on the weight given to the costs; solutions (land allocations) for different weightings fall along a trade-offs curve (figure 1a). Using trade-offs curves, a set of areas can be identified that achieves a given level of biodiversity representation/protection, with minimum opportunity costs. Trade-offs space also can be used to explore scenarios – for example, whether a fixed allocation of areas to one land use means that the best trade-offs curve for the region is now much worse.

Clearly, if net benefits for society are important, biodiversity assessment cannot occur without other sectoral factors "on the table". Decision-making about land allocations that ignores such factors not only implies lower net benefits at that time, but also may imply that it is no longer possible to achieve high net benefits in the future. Such constraints might arise through previous land use decisions or through impacts such as climate change.

Constraints also may be a consequence of loss of degraded land to both biodiversity protection and other land use opportunities, fixed protected areas with high opportunity cost but low biodiversity representation, or fixed production areas with high biodiversity loss but low production opportunity. With such constraints, the new trade-offs curve of best-possible solutions moves away from the optimum (implying lower net benefits; figure 1a), and there is a reduction in "regional sustainability" – the degree to which the region has achieved its capacity for finding a balance among competing needs of society.

Allocation of different areas to different land uses so as to maximise net benefits is only one key to finding a balance between competing needs of society. The search for net benefits also recognises cases where two or more "services" can be met in a single area. Net benefits to society, a better balance between potentially conflicting goals, can therefore be achieved on occasions when it is found that the different land uses are not so much in conflict. "Partial protection" of biodiversity – assigning land uses to areas that provide at least partial protection of biodiversity while not forgoing other land use opportunities – leads to a trade-offs curve providing greater net benefits. In contrast, higher curves in tradeoffs space (offering lower net benefits; figure 1a) arise when there are fewer opportunities for such partial protection.

When partial protection of biodiversity is credited to other land uses, it can be taken into account in calculating complementarity values for regional planning. Eco-forestry, for example, may provide at least partial protection of biodiversity. When we take into account those contributions, clearly there is then reduced pressure on other areas to contribute to biodiversity goals, and so there is greater opportunity for other land use opportunities.

In regional planning, we can identify specific areas where most is gained by implementing a form of development that achieves partial protection. Economic incentives for such sympathetic management then may be applied to those areas. Such analyses are currently being explored in our study applying trade-offs in regional tourism planning in Douglas Shire, North Queensland, Australia.

Trade-offs Studies in Australia and Papua New Guinea

We have been working on cross-sectoral trade-offs approaches, incorporating biodiversity and opportunity costs, since our early case study in New South Wales, Australia. That study demonstrated advantages of trading-off biodiversity and other sectoral needs of society at the regional level (figure 1b). While not knowing any "correct" weighting for various costs, sensitivity analysis revealed which areas were selected for protection no matter what the weight assigned to opportunity costs – and which areas were never selected. That study also highlighted another important aspect of trade-offs.

The optimal solution had low overlap with one that would have achieved the same biodiversity target ignoring costs (figure 1b), suggesting that consideration of costs cannot simply be an add-on consideration to refine land allocations. The New South Wales study represented the first integration of general opportunity costs into conservation priority-setting algorithms, and the same basic approach has now been adopted elsewhere.

Our more recent trade-offs project identified candidate areas for biodiversity protection in Papua New Guinea (PNG), providing an ongoing evaluation framework for moving towards a country-wide conservation goal, while at the same time providing opportunities to alter the priority area set in the light of new knowledge, changes in land use, and/or changes in economic and social conditions.

We applied a new approach to percentage-based conservation targets in PNG, based on trade-offs. The maximum diversity that could be protected by an unconstrained 10% of total area became the working biodiversity target. Reaching that same biodiversity target then required more than 10% of the total area, because of trade-offs involving constraints and opportunity costs. The subsequent satisfaction of the 10%-based target in a low-cost proposed protected set covering 16.8% of PNG corresponded to relatively high net benefits (figure 2).

Achieving a biodiversity protection target with minimum opportunity cost was an important outcome in PNG, given that biodiversity values overlap with forestry production values, and high forgone forestry opportunities would mean significant losses to land owners and the government. In the PNG study, the same level of biodiversity representation conceivably could have corresponded to a small or large overlap with the desirable areas for forestry production. Clearly, achieving net benefits means that evaluations and planning cannot be carried out without taking into account such factors as timber production, agriculture, population centres, carbon sequestration, and the economics of government levies, incentives and subsidies.

The PNG study demonstrated that complementarity is not just about selecting a set of priority protected areas, but about a new 'biodiversity economics' relating to offsets, levies, subsidies, incentives and other economic instruments. Future scenario development in PNG therefore will focus on biodiversity targets, land use constraints, timber plans, population issues, scope for levies, offsets markets, and subsidies.

Regional sustainability was addressed at two interacting levels in the PNG study. First, the planning approach minimised forgone timber opportunities for a given biodiversity protection level. Second, the analysis framework addressed the potential for eco-forestry as potentially replacing intensive logging, so that areas producing logging income could contribute to biodiversity protection.

The PNG study, however, did not credit allocations of partial protection of biodiversity towards achieving the target. Future trade-offs may be achieved even more effectively when production lands are credited in the allocation process to at least partial protection of biodiversity (when they do actually make such a contribution). Our current approach to integration of such credits uses probabilistic strategies for expressing biodiversity persistence in a planning framework.

Because the PNG study did not incorporate partial protection allocations and, overall, did not capture the full breadth of trade-offs space and scenarios analysis, we will present an hypothetical example to illustrate important aspects of trade-offs. We will present example analyses, using our trade-offs software, TARGET, to highlight the role of biodiversity trade-offs in providing a framework for sustainability, both across sectors and across spatial scales.

Those examples then set the stage for our consideration of the role of such trade-offs in several new international initiatives. Now, we will examine two of these, the millennium ecosystem assessment and the critical ecosystems (CE) or "hotspots" programme. The need to move beyond single sector assessments is a central rationale for the MA, which is concerned with trade-offs among "ecosystem services" that include such things as water quality, food production, and conservation of biodiversity option values.

The CE has a focus on conserving the key biodiversity hotspot areas of the world, with a prioritisation strategy that takes into account the need to minimise opportunity costs. Both the MA and CE so far appear to have no explicit framework that incorporates the sorts of trade-offs used in our case studies, where balance is achieved by allocation among areas and crediting partial protection. For both programmes, we will argue that our trade-offs framework can provide a natural linkage between local, regional and global planning levels. A balance within individual areas feeds into the assessment of trade-offs at the regional level – and the potential of that region for trade-offs may make it more deserving of high global priority for conservation efforts.

Target Using Trade-offs Analyses

Target is one module of the diversity software package. It is also part of the BioRap toolbox, and was used in the PNG trade-offs analyses. The biodiversity trade-offs strategies used in target were originally developed in one of the other diversity modules. Target not only provides for analyses, as in the PNG study, that seek to achieve a biodiversity protection target at least cost, but also provides procedures to explore scenarios in trade-offs spaces (figure 1a).

Target processes locationally referenced data and links to spatial mapping software. Target assumes that the areas in a region are described as containing one or more different biodiversity "attributes". Attributes may be species or other surrogate information for biodiversity. For example, the attributes at the coarsest level of resolution might be forest types, with further definition of finer-resolution attributes describing variation within forest types.

The finer-resolution attributes might correspond to different species, or might be other abiotic descriptions, such as soil types. Within each area, each attribute also has some quantitative value associated with it – this value might, for example, correspond to the total number of hectares of that forest type within that area.

Input to target consists of this list of geographic areas, each containing a set of attributes found in the area, with some quantitative value associated with each. Also, a nominated degree of representation for each attribute is provided as input. The first step in using target involves setting that target level for representation for all biodiversity attributes. These target levels may be derived from consideration of standard regional biodiversity "targets". For example, the Caracas Action Plan, which is linked to the protected-areas elements of the Convention on Biological Diversity, proposed as a broad target that 10% of each biome be represented in a protected areas system.

Target implements one strategy for applying 10% (or similar) targets so that they can more effectively be used as comparative performance indicators among countries or regions. A benchmark analysis may be carried out, as in the PNG study to determine how much heterogeneity (how many attributes) could be sampled (perhaps to some predefined viability level) under an assumption that any 10% of the region can be selected. This level of heterogeneity becomes the new effective biodiversity target.

The target amount for each attribute then may be set, as it was in the PNG study, to a single viable representation of each attribute. Subsequent analyses ask how this target can be reached in the face of real constraints and opportunity costs. Basic target analyses search for a set of candidate protected areas that achieves nominated target levels of representation of all the attributes, but with a minimum opportunity cost. These opportunity costs of biodiversity protection often will correspond to estimates of the suitability of the areas for forestry production or other land uses.

When costs are taken into account, the relative weight given to these costs, relative to biodiversity representation, will influence the outcome of the allocation procedure. An area is justified for protection if and only if its complementarity value exceeds its weighted cost. Target iteratively adds and deletes areas from its list of priority protected areas so as to approach the nominated target levels of representation.

During the course of analyses, the software repeatedly calculates a new complementarity value for any given area, reflecting changes in the degree of additional attribute representation/ protection that the area can contribute to the list of protected areas. When trade-offs are used, target attempts to balance this complementarity contribution against the specified costs of protection. The area which is added to the selected priority set, at any stage, is the one which has the greatest difference between complementarity and cost.

The user can take advantage of target capabilities to extend and modify the simple search provided by the basic algorithm. One approach can use alternative random starts. Another approach can begin with a high weighting on costs, such that targets are not met, and the reading in of this partial result into a subsequent analysis with lower weight on costs. This strategy can be applied iteratively until the target is met. Similar iterative approaches might initially mask out some areas, giving preference to others until later iterations.

In practice, we have found these strategies useful in identifying optimal solutions. We note that the basic step-wise trade-offs algorithm contrasting costs and complementarity has now been implemented in other software (e.g. "WorldMap"), but the limited search options means that least-cost solutions are less likely to be identified. When costs are given high weight in target analyses, some initial biodiversity targets are not reached. A land cover type, for example, may have a lower level of representation because protection of that type generally implies higher opportunity costs.

Land cover types that are more heterogeneous (have a greater number of different attributes describing variation within the type) may justify greater representation, particularly in the presence of competing land use demands. In addition to costs, the other factor that properly should influence the amount of area needed for representation/ protection of an attribute is the degree to which areas of that type are likely to persist in the absence of formal protection. Quite extensive attributes may require a relatively small per centage of their total area in formal protection, because the extent of coverage helps ensure overall regional persistence of that type.

Each biodiversity attribute in a given area may have some assumed degree of persistence in the absence of any new land-use allocation for the area, and some different degree of persistence if the area is allocated to a particular land use (e.g. formal protection or sympathetic management). Examples of degrees of persistence values, based on probabilities, are shown in the PNG example. Target uses one simple strategy in which partial protection in the absence of action/re-allocation can be taken into account through modification of targets, and there is some further partial protection or persistence value assignable for each attribute if the given area is allocated to protection.

The usual quantitative values in the input data files, associated with each attribute in each area, are interpreted as indicating the degree of persistence of the attribute if that area were allocated to protection. The gain in total regional degree of persistence for a given attribute is a function of that attribute's individual persistence values in the set of protected areas.

When the values for all areas are negative log transforms of probabilities of extinction, then summing these values provides a (negative log transform) of the overall regional probability of extinction of the attribute. The associated regional target for the attribute may be a 0.999 probability of persistence, equivalent to a 0.001 probability of extinction (a negative log10 transform value of 3).

Millennium Assessment (MA)

The MA is a new international programme of ecosystem assessments, organised and supported by UNEP, IUCN, World Bank, WRI, and others with the "goal of improving management of ecosystems (at all scales) by providing information to decision-makers about the condition or "health of ecosystems, consequences of ecosystem change, and options for response (policy, legislative, technological, etc.)". Health here is seen as the capacity to supply goods and services to society. Daily refers to these "ecosystem services" as "the conditions and processes through which natural ecosystems, and the species that make them up, sustain and fulfill human life".

The MA identifies trade-offs among services as a key issue, arguing that "the challenge of meeting the human needs for ecosystem goods and services is so great that trade-offs have become the rule" and that "it remains to be seen whether the capacity of the system to provide the combination of the services is optimised". The challenge therefore is to learn as much as possible about interlinkages among services.

Further, "only be looking at the entire array of goods and services provided by ecosystems can wise decisions be made that address the interlinkages among them". "The MA is thus concerned with examining how changes in ecosystems affect access to basic needs such as food, fuel, fiber, water; how they affect local climate and risks associated with droughts, storms, floods; how they affect human health and human economies; and how they affect less tangible cultural, moral, aesthetic, and ethical concerns".

Conservation of Biodiversity and Other Specific Services

The conservation of biodiversity is recognised as a key ecosystem service. However, given the wide range of different services competing for attention in the MA, it is natural to ask whether biodiversity conservation will be ignored in practice. While the argument can be made that protecting other services protects biodiversity, actions to maintain specific species – the "biospecifics" associated with specific services –does not necessarily maintain overall biodiversity as well.

Biodiversity conservation must be given some explicit weight relative to other services for effective tradeoffs to be achieved. The process of trading-off services in the market-based economic framework of the MA also may lead to a neglect of biodiversity and its option values. In the context of the MA, Ayensu *et al* call for new information on the economic value of "non-marketed" goods and services produced by ecosystems, noting that economic values "may be significant to management decisions". Because the ecosystem services approach looks for economic values, the option values of biodiversity may be hard to quantify as part of the same assessments.

Conserving the option value of biodiversity may be an acknowledged ecosystem service, but there is a danger that it will be left out of assessments as too hard to quantify in terms of economic value. In spite of these concerns, the focus on trade-offs in the MA is potentially quite compatible with non-economic valuations as well. For example, Daily states, "this forecasting exercise will illuminate the tradeoffs among different ecosystem variables, and therefore among different mixtures of ecosystem goods and services (which may be measured in contrasting units such as mass of carbon or diversity of species)".

In accord with this, we suggest that a land-use allocation trade-offs space, with emphasis on complementarity among areas, can play an important role in the MA. As the PNG study illustrates, there can be a mix of markets together with "top-down" imposed tradeoffs that set biodiversity targets based on quantification of biodiversity option values.

Changes of Complementarity

In exploring scenarios of human-induced change on ecosystems and consequent impacts on trade-offs among services, it is important to consider changes in biodiversity complementarity values. We have seen that interdependencies among different constraints/factors, as in the target examples, account for changes in complementarity values. For example, areas 17 and 18 had non-zero complementarity only when partial protection in areas 6 and 7 was incorporated into the regional plan for biodiversity protection. In the PNG study, changes of complementarity value under different scenarios affected expectations about carbon offsets, levies and other services.

The key role of biodiversity complementarity in tradeoffs among services needs to be considered in the context of the MA's stated perspectives on biodiversity. The MA planning documents describe biodiversity information largely as species-richness or total diversity. For example, references are made to the need for richness maps for different regions and ecosystems. However, regional trade-offs must compare an area's "marginal", context-dependent, biodiversity contribution to other services.

Similarly, the MA asks: "Where is the condition of the ecosystem to maintain biodiversity in good condition and where is it declining? What do we know about the impact of habitat change on species diversity and how does this differ in different regions and taxonomic groups? What do we know about the impact of fragmentation on species diversity?". We argue that impacts on complementarity values are perhaps more critical to trade-offs.

The MA's focus on diversity or total richness is revealed also in the call for "an additional analysis of the state of our scientific understanding of whether or not patterns of species richness in known taxonomic groups match patterns in poorly known groups". The trade-offs framework, in contrast, focuses on a quite different form of prediction, asking whether the complementarity value for one taxonomic group predicts the complementarity of other groups.

Other discussions of the MA have given less emphasis to total diversity. Daily's proposed framework for trading-off ecosystem services highlights marginal gains: "The evaluation of the tradeoffs currently facing society, however, requires estimating the marginal value of ecosystem services to determine the costs of losing-or benefits of preserving – a given amount or quality of service". We endorse this perspective as compatible with the role of complementarity in trade-offs, but with one caveat.

In our biodiversity tradeoffs framework, the quantities being tradedoffs are more general than ecosystem services. Often the opportunity costs are not a good or service: in PNG a cost of biodiversity protection is the "transaction" cost, equal, for example, to costs of administration.

The MA's emphasis on total species diversity (rather than complementarity) is partly a natural consequence of seeing diversity as a driving factor influencing other services. In considering impacts of changes in biodiversity, the MA asks: "How does change in species diversity affect the biological productivity of ecosystems and in particular the production of fish, food, fiber and other products directly consumed by people?" A "balance" among services within an area is achieved if such an interdependency exists – maintaining biodiversity implies maintenance of those other services as well.

More generally, those considerations raise issues of partial protection, where biodiversity conservation and maintenance of other services go hand-in-hand. The broad scope of the problem of balancing various services within and among areas suggests to us that the general property of our trade-offs framework that is of most potential use for the MA is its unified view of tradeoffs over different scales. "Partial protection" of biodi-versity within an area feeds into "regional sustainability" assessment at higher scales. The MA, which has a strong focus already on finding a balance among "ecosystem services" *within* areas, will benefit also from trade-offs *among* areas as part of its scenarios analyses.

Critical Ecosystems (CE)

The CE programme is based in part on Conservation International's programmes identifying 25 global hotspots, based on places having high estimated levels of endemicity. The CE hotspots approach excludes trading-off biodiversity with opportunity costs as a strategy for prioritising among regions: "We should also emphasise here that the biological criteria are always the first cut, the first layer of analysis, in determining the priority status of a particular region.

We believe that it is dangerous and misleading to mix threat and biological criteria in the first step of analysis, and even more confusing when social, economic, and even political feasibility criteria are mixed into this first level. It is important always to keep in mind that a biodiversity priority-setting exercise must focus first and foremost on the biological, and that other criteria such as threat, social and economic factors, political will, feasibility, and the like should be introduced in subsequent layers of analysis. Indeed, we believe that these are most useful in the project design phase, when such factors become particularly relevant".

While we strongly advocate trade-offs for priority setting, we do agree, in part, with this position. At the level of prioritising among regions, endemicity is useful as a key factor. First, endemism contributions to complementarity values can determine the "must-have" areas in any trade-offs analysis that is to represent all taxa. Those areas are given priority no matter what the cost. Further, the suggested application of the socioeconomic factors only at the "project design" phase can be interpreted to encompass trade-offs to determine which areas *within* the region are to be protected for biodiversity conservation.

The hotspots approach recently has been justified as the most cost effective way to protect biodiversity. Mace *et al* argued that the claim of cost-efficiency was suspect without application of tradeoffs and other approaches to priority setting. However, simple application of trade-offs selection procedures at the *among-regions* level does not capture the proper role for trade-offs in priority setting. Trade-offs, given their key role *within regions*, may influence priorities *among* regions simply because we want to know where they are most urgently needed.

The CE programme, in spite of the biology-only criteria quoted above, acknowledges such a consideration: "The three major wilderness areas of the Congo, the Amazon, and New Guinea and associated forests in insular Southeast Asia not only contain a very large fraction of the world's species, but they are among the very places where natural process unfold on something like their natural scale. Their protection has to be a major priority. These areas suffer from a variety of threats that include logging, mining, the extraction of oil and gas, and the conversion to monoculture agriculture, such as soybeans". Thus, the CE programme appears compatible with trade-offs within a region, possibly with some higher-level priority given to those regions where the trade-offs within are most needed.

This idealised view also can be reconciled in practice with trade-offs. The trade-offs framework may play an important role in either (i) justifying why all the remaining intact habitat must be protected, or alternatively (ii) showing how some areas can be allocated to other uses so as to reduce overall costs to society in terms of forgone opportunities.

In the first case, a 10% (or other) conservation target may be applied as it was in the PNG study. Meeting the target means protecting biodiversity within the region to the extent that it could have been accomplished with an unconstrained 10% of the area. If land clearance/disturbance is extensive, then all remaining areas may require protection to meet that biodiversity target. In the second case, some trade-offs are still possible in achieving the conservation target, and opportunity costs therefore can be minimised through a combination of land allocation and partial protection, where multiple ecosystem services are accommodated within the same region and sometimes even within the same area.

Simply assuming all undisturbed areas within a hotspot deserve protection could imply an unduly great opportunity cost of conservation. When trade-offs are possible, there can be a large penalty to society to ignore them (figure 1a). In the PNG study, hypothetical analyses based on biodiversity protection without regard to costs created unnecessarily low net benefits.

Given that PNG is part of the CE priority tropical wilderness areas for conservation, it is interesting also that PNG is regarded as having low costs: "Whereas the hotspots consist of heavily exploited and often highly fragmented ecosystems greatly reduced in original extent (usually between 4% and 25% remaining), the major tropical wilderness areas are still largely intact (over 75% of original vegetation cover remaining) and have very low human population density (less than 5 people/km^2).

Myers had touched on a concept of this kind in his first hotspots paper, briefly referring to what he called "good news areas". Further, "since they are still under far less human pressure than the threatened hotspots (although the pressures on them are increasing rapidly), the "opportunity cost" of conservation is much lower in these areas, i.e. large-scale conservation set-asides can be achieved at far lower financial cost than in the areas where little remains". Thus, low population density and large areas "intact" are seen as implying low opportunity costs of conservation. We see PNG as indeed not devastated, but not a "good news" area either. In truth, PNG can imply low realised opportunity costs or quite high realised opportunity costs, depending on whether a balance is found through biodiversity planning based on trade-offs.

References

Bowen B. W. (1999). "Preserving genes, species, or ecosystems? Healing the fractured foundations of conservation policy"(PDF). *Molecular Ecology* 8: S5–S10.

Brooks T. M., Mittermeier R. A., Gerlach J., Hoffmann M., Lamoreux J. F., Mittermeier C. G., Pilgrim J. D., Rodrigues A. S. L. (2006). "Global Biodiversity Conservation Priorities". *Science* 313 (5783): 58.

Kareiva P., Marvier M. (2003). "Conserving Biodiversity Coldspots" (PDF). *American Scientist* 91 (4): 344–351.

McCallum M. L. (2008). "Amphibian Decline or Extinction? Current Declines Dwarf Background Extinction Rate" (PDF).*Journal of Herpetology* 41 (3): 483–491.

Myers N., Mittermeier R. A., Mittermeier C. G., Kent J. (2000). "Biodiversity hotspots for conservation priorities". *Nature* 403: 853–858.

Primack, Richard B. (2004). *A primer of Conservation Biology*. Sunderland, Mass: Sinauer Associates.

8

Convention on Biological Diversity

The Convention on Biological Diversity (CBD), known informally as the Biodiversity Convention, is an international legally binding treaty. The Convention has three main goals:

— conservation of biological diversity (or biodiversity);

— sustainable use of its components; and

— fair and equitable sharing of benefits arising from genetic resources

In other words, its objective is to develop national strategies for the conservation and sustainable use of biological diversity. It is often seen as the key document regarding sustainable development.

The Convention was opened for signature at the Earth Summit in Rio de Janeiro on 5 June 1992 and entered into force on 29 December 1993.

2010 was the International Year of Biodiversity. The Secretariat of the Convention on Biological Diversity is the focal point for the International Year of Biodiversity. At the 2010 10th Conference of Parties (COP) to the Convention on Biological Diversity in October in Nagoya, Japan, the Nagoya Protocol was adopted. On 22 December 2010, the UN declared the period from 2011 to 2020 as the UN-Decade on Biodiversity. They, hence, followed a recommendation of the CBD signatories during COP10 at Nagoya in October 2010.

The convention recognized for the first time in international law that the conservation of biological diversity is "a common concern of humankind" and is an integral part of the development process. The agreement covers all ecosystems, species, and genetic resources. It links traditional conservation efforts to the economic goal of using biological resources sustainably. It sets principles for the fair and equitable sharing of the benefits arising from the use of genetic resources, notably those destined for commercial use. It also covers the rapidly expanding field of biotechnology through its Cartagena Protocol on Biosafety, addressing technology development and transfer, benefit-sharing and biosafety issues. Importantly, the Convention is legally binding; countries that join it ('Parties') are obliged to implement its provisions.

The convention reminds decision-makers that natural resources are not infinite and sets out a philosophy of sustainable use. While past conservation efforts were aimed at protecting particular species and habitats, the Convention recognizes that ecosystems, species and genes must be used for the benefit of humans. However, this should be done in a way and at a rate that does not lead to the long-term decline of biological diversity.

The convention also offers decision-makers guidance based on the precautionary principle that where there is a threat of significant reduction or loss of biological diversity, lack of full scientific certainty should not be used as a reason for postponing measures to avoid or minimize such a threat. The Convention acknowledges that substantial investments are required to conserve biological diversity. It argues, however, that conservation will bring us significant environmental, economic and social benefits in return.

The fulltext of the convension is given below.

Preamble

The Contracting Parties,

Conscious of the intrinsic value of biological diversity and of the ecological, genetic, social, economic, scientific, educational, cultural, recreational and aesthetic values of biological diversity and its components,

Conscious also of the importance of biological diversity for evolution and for maintaining life sustaining systems of the biosphere,

Affirming that the conservation of biological diversity is a common concern of humankind,

Reaffirming that States have sovereign rights over their own biological resources,

Reaffirming also that States are responsible for conserving their biological diversity and for using their biological resources in a sustainable manner,

Concerned that biological diversity is being significantly reduced by certain human activities,

Aware of the general lack of information and knowledge regarding biological diversity and of the urgent need to develop scientific, technical and institutional capacities to provide the basic understanding upon which to plan and implement appropriate measures,

Noting that it is vital to anticipate, prevent and attack the causes of significant reduction or loss of biological diversity at source, Noting also that where there is a threat of significant reduction or loss of biological diversity, lack of full scientific certainty should not be used as a reason for postponing measures to avoid or minimize such a threat,

Noting further that the fundamental requirement for the conservation of biological diversity is the in-situ conservation of ecosystems and natural habitats and the maintenance and recovery of viable populations of species in their natural surroundings, Noting further that ex-situ measures, preferably in the country of origin, also have an important role to play,

Recognizing the close and traditional dependence of many indigenous and local communities embodying traditional lifestyles on biological resources, and the desirability of sharing equitably benefits arising from the use of traditional knowledge, innovations and practices relevant to the conservation of biological diversity and the sustainable use of its components,

Recognizing also the vital role that women play in the conservation and sustainable use of biological diversity and affirming the need for the full participation of women at all levels of policymaking and implementation for biological diversity conservation,

Stressing the importance of, and the need to promote, international, regional and global cooperation among States and intergovernmental organizations and the non-governmental sector for the conservation of biological diversity and the sustainable use of its components,

Acknowledging that the provision of new and additional financial resources and appropriate access to relevant technologies can be expected to make a substantial difference in the world's ability to address the loss of biological diversity,

Acknowledging further that special provision is required to meet the needs of developing countries, including the provision of new and additional financial resources and appropriate access to relevant technologies,

Noting in this regard the special conditions of the least developed countries and small island States,

Acknowledging that substantial investments are required to conserve biological diversity and that there is the expectation of a broad range of environmental, economic and social benefits from those investments,

Recognizing that economic and social development and poverty eradication are the first and overriding priorities of developing countries,

Aware that conservation and sustainable use of biological diversity is of critical importance for meeting the food, health and other needs of the growing world population, for which purpose access to and sharing of both genetic resources and technologies are essential,

Noting that, ultimately, the conservation and sustainable use of biological diversity will strengthen friendly relations among States and contribute to peace for humankind,

Desiring to enhance and complement existing international arrangements for the conservation of biological diversity and sustainable use of its components, and

Determined to conserve and sustainably use biological diversity for the benefit of present and future generations, Have agreed as follows:

Article 1. Objectives

The objectives of this Convention, to be pursued in accordance with its relevant provisions, are the conservation of biological diversity, the sustainable use of its components and the fair and equitable sharing of the benefits arising out of the utilization of genetic resources, including by appropriate access to genetic resources and by appropriate transfer of relevant technologies, taking into account all rights over those resources and to technologies, and by appropriate funding.

Article 2. Use of Terms

For the purposes of this Convention:

"Biological diversity" means the variability among living organisms from all sources including, inter alia, terrestrial, marine and other aquatic ecosystems and the ecological complexes of which they are part; this includes diversity within species, between species and of ecosystems.

"Biological resources" includes genetic resources, organisms or parts thereof, populations, or any other biotic component of ecosystems with actual or potential use or value for humanity. "Biotechnology" means any technological application that uses biological systems, living organisms, or derivatives thereof, to make or modify products or processes for specific use. "Country of origin of genetic resources" means the country which possesses those genetic resources in in-situ conditions. "Country providing genetic resources" means the country supplying genetic resources collected from in-situ sources, including populations of both wild and domesticated species, or taken from ex-situ sources, which may or may not have originated in that country. "Domesticated or cultivated species" means species in which the evolutionary process has been influenced by humans to meet their needs. "Ecosystem" means a dynamic complex of plant, animal and micro-organism communities and their non-living environment interacting as a functional unit.

"Ex-situ conservation" means the conservation of components of biological diversity outside their natural habitats. "Genetic material" means any material of plant, animal, microbial or other origin containing functional units of heredity. "Genetic resources" means genetic material of actual or potential value.

"Habitat" means the place or type of site where an organism or population naturally occurs.

"In-situ conditions" means conditions where genetic resources exist within ecosystems and natural habitats, and, in the case of domesticated or cultivated species, in the surroundings where they have developed their distinctive properties.

"In-situ conservation" means the conservation of ecosystems and natural habitats and the maintenance and recovery of viable populations of species in their natural surroundings and, in the case of domesticated or cultivated species, in the surroundings where they have developed their distinctive properties.

"Protected area" means a geographically defined area which is designated or regulated and managed to achieve specific conservation objectives.

"Regional economic integration organization" means an organization constituted by sovereign States of a given region, to which its member States have transferred competence in respect of matters governed by this Convention and which has been duly authorized, in accordance with its internal procedures, to sign, ratify, accept, approve or accede to it.

"Sustainable use" means the use of components of biological diversity in a way and at a rate that does not lead to the long-term decline of biological diversity, thereby maintaining its potential to meet the needs and aspirations of present and future generations. "Technology" includes biotechnology.

Article 3. Principle

States have, in accordance with the Charter of the United Nations and the principles of international law, the sovereign right to exploit their own resources pursuant to their own environmental policies, and the responsibility to ensure that activities within their jurisdiction or control do not cause damage to the environment of other States or of areas beyond the limits of national jurisdiction.

Article 4. Jurisdictional Scope

Subject to the rights of other States, and except as otherwise expressly provided in this Convention, the provisions of this Convention apply, in relation to each Contracting Party:

(a) In the case of components of biological diversity, in areas within the limits of its national jurisdiction; and

(b) In the case of processes and activities, regardless of where their effects occur, carried out under its jurisdiction or control, within the area of its national jurisdiction or beyond the limits of national jurisdiction.

Article 5. Cooperation

Each Contracting Party shall, as far as possible and as appropriate, cooperate with other Contracting Parties, directly or, where appropriate, through competent international organizations, in respect of areas beyond national jurisdiction and on other matters of mutual interest, for the conservation and sustainable use of biological diversity.

Article 6. General Measures for Conservation and Sustainable Use

Each Contracting Party shall, in accordance with its particular conditions and capabilities:

(a) Develop national strategies, plans or programmes for the conservation and sustainable use of biological diversity or adapt for this purpose existing strategies, plans or programmes which shall reflect, inter alia, the measures set out in this Convention relevant to the Contracting Party concerned; and

(b) Integrate, as far as possible and as appropriate, the conservation and sustainable use of biological diversity into relevant sectoral or cross-sectoral plans, programmes and policies.

Article 7. Identification and Monitoring

Each Contracting Party shall, as far as possible and as appropriate, in particular for the purposes of Articles 8 to 10:

(a) Identify components of biological diversity important for its conservation and sustainable use having regard to the indicative list of categories set down in Annex I;

(b) Monitor, through sampling and other techniques, the components of biological diversity identified pursuant to subparagraph

(a) above, paying particular attention to those requiring urgent conservation measures and those which offer the greatest potential for sustainable use;

(c) Identify processes and categories of activities which have or are likely to have significant adverse impacts on the conservation and sustainable use of biological diversity, and monitor their effects through sampling and other techniques; and

(d) Maintain and organize, by any mechanism data, derived from identification and monitoring activities pursuant to subparagraphs (a),

(b) and (c) above.

Article 8. In-situ Conservation

Each Contracting Party shall, as far as possible and as appropriate:

(a) Establish a system of protected areas or areas where special measures need to be taken to conserve biological diversity;

(b) Develop, where necessary, guidelines for the selection, establishment and management of protected areas or areas where special measures need to be taken to conserve biological diversity;

(c) Regulate or manage biological resources important for the conservation of biological diversity whether within or outside protected areas, with a view to ensuring their conservation and sustainable use;

(d) Promote the protection of ecosystems, natural habitats and the maintenance of viable populations of species in natural surroundings;

(e) Promote environmentally sound and sustainable development in areas adjacent to protected areas with a view to furthering protection of these areas;

(f) Rehabilitate and restore degraded ecosystems and promote the recovery of threatened species, inter alia, through the development and implementation of plans or other management strategies;

(g) Establish or maintain means to regulate, manage or control the risks associated with the use and release of living modified organisms resulting from biotechnology which are likely to have adverse environmental impacts that could affect the conservation and sustainable use of biological diversity, taking also into account the risks to human health;

(h) Prevent the introduction of, control or eradicate those alien species which threaten ecosystems, habitats or species;

(i) Endeavour to provide the conditions needed for compatibility between present uses and the conservation of biological diversity and the sustainable use of its components;

(j) Subject to its national legislation, respect, preserve and maintain knowledge, innovations and practices of indigenous and local communities embodying traditional lifestyles relevant for the conservation and sustainable use of biological diversity and promote their wider application with the approval and involvement of the holders of such knowledge, innovations and practices and encourage the equitable sharing of the benefits arising from the utilization of such knowledge, innovations and practices;

(k) Develop or maintain necessary legislation and/or other regulatory provisions for the protection of threatened species and populations;

(l) Where a significant adverse effect on biological diversity has been determined pursuant to Article 7, regulate or manage the relevant processes and categories of activities; and

(m) Cooperate in providing financial and other support for insitu conservation outlined in subparagraphs (a) to (l) above, particularly to developing countries.

Article 9. Ex-situ Conservation

Each Contracting Party shall, as far as possible and as appropriate, and predominantly for the purpose of complementing in-situ measures:

(a) Adopt measures for the ex-situ conservation of components of biological diversity, preferably in the country of origin of such components;

(b) Establish and maintain facilities for ex-situ conservation of and research on plants, animals and micro-organisms, preferably in the country of origin of genetic resources;

(c) Adopt measures for the recovery and rehabilitation of threatened species and for their reintroduction into their natural habitats under appropriate conditions;

(d) Regulate and manage collection of biological resources from natural habitats for ex-situ conservation purposes so as not to threaten ecosystems and in-situ populations of species, except where special temporary ex-situ measures are required under subparagraph (c) above; and

(e) Cooperate in providing financial and other support for exsitu conservation outlined in subparagraphs (a) to (d) above and in the establishment and maintenance of ex-situ conservation facilities in developing countries.

Article 10. Sustainable Use of Components of Biological Diversity

Each Contracting Party shall, as far as possible and as appropriate:

(a) Integrate consideration of the conservation and sustainable use of biological resources into national decision-making;

(b) Adopt measures relating to the use of biological resources to avoid or minimize adverse impacts on biological diversity;

(c) Protect and encourage customary use of biological resources in accordance with traditional cultural practices that are compatible with conservation or sustainable use requirements;

(d) Support local populations to develop and implement remedial action in degraded areas where biological diversity has been reduced; and

(e) Encourage cooperation between its governmental authorities and its private sector in developing methods for sustainable use of biological resources.

Article 11. Incentive Measures

Each Contracting Party shall, as far as possible and as appropriate, adopt economically and socially sound measures that act as incentives for the conservation and sustainable use of components of biological diversity.

Article 12. Research and Training

The Contracting Parties, taking into account the special needs of developing countries, shall:

(a) Establish and maintain programmes for scientific and technical education and training in measures for the identification, conservation and sustainable use of biological diversity and its components and provide support for such education and training for the specific needs of developing countries;

(b) Promote and encourage research which contributes to the conservation and sustainable use of biological diversity, particularly in developing countries, inter alia, in accordance with decisions of the Conference of the Parties taken in consequence of recommendations of the Subsidiary Body on Scientific, Technical and Technological Advice; and

(c) In keeping with the provisions of Articles 16, 18 and 20, promote and cooperate in the use of scientific advances in biological diversity research in developing methods for conservation and sustainable use of biological resources.

Article 13. Public Education and Awareness

The Contracting Parties shall:

(a) Promote and encourage understanding of the importance of, and the measures required for, the conservation of biological diversity, as well as its propagation through media, and the inclusion of these topics in educational programmes; and

(b) Cooperate, as appropriate, with other States and international organizations in developing educational and public awareness programmes, with respect to conservation and sustainable use of biological diversity.

Article 14. Impact Assessment and Minimizing Adverse Impacts

1. Each Contracting Party, as far as possible and as appropriate, shall:

   (a) Introduce appropriate procedures requiring environmental impact assessment of its proposed projects that are likely to have significant adverse effects on biological diversity with a view to avoiding or minimizing such effects and, where appropriate, allow for public participation in such procedures;

   (b) Introduce appropriate arrangements to ensure that the environmental consequences of its programmes and policies that are likely to have significant adverse impacts on biological diversity are duly taken into account;

   (c) Promote, on the basis of reciprocity, notification, exchange of information and consultation on activities under their jurisdiction or control which are likely to significantly affect adversely the biological diversity of other States or areas beyond the limits of national jurisdiction, by encouraging the conclusion of bilateral, regional or multilateral arrangements, as appropriate;

   (d) In the case of imminent or grave danger or damage, originating under its jurisdiction or control, to biological diversity within the area under jurisdiction of other States or in areas beyond the limits of national jurisdiction, notify immediately the potentially affected States of such danger or damage, as well as initiate action to prevent or minimize such danger or damage; and

   (e) Promote national arrangements for emergency responses to activities or events, whether caused naturally or otherwise, which present a grave and imminent danger to biological diversity and encourage international cooperation to supplement such national efforts and, where appropriate and agreed by the States or regional economic integration organizations concerned, to establish joint contingency plans.

2. The Conference of the Parties shall examine, on the basis of studies to be carried out, the issue of liability and redress, including restoration and compensation, for damage to biological diversity, except where such liability is a purely internal matter.

Article 15. Access to Genetic Resources

1. Recognizing the sovereign rights of States over their natural resources, the authority to determine access to genetic resources rests with the national governments and is subject to national legislation.
2. Each Contracting Party shall endeavour to create conditions to facilitate access to genetic resources for environmentally sound uses by other Contracting Parties and not to impose restrictions that run counter to the objectives of this Convention.
3. For the purpose of this Convention, the genetic resources being provided by a Contracting Party, as referred to in this Article and Articles 16 and 19, are only those that are provided by Contracting Parties that are countries of origin of such resources or by the Parties that have acquired the genetic resources in accordance with this Convention.
4. Access, where granted, shall be on mutually agreed terms and subject to the provisions of this Article.
5. Access to genetic resources shall be subject to prior informed consent of the Contracting Party providing such resources, unless otherwise determined by that Party.
6. Each Contracting Party shall endeavour to develop and carry out scientific research based on genetic resources provided by other Contracting Parties with the full participation of, and where possible in, such Contracting Parties.
7. Each Contracting Party shall take legislative, administrative or policy measures, as appropriate, and in accordance with Articles 16 and 19 and, where necessary, through the financial mechanism established by

Articles 20 and 21 with the aim of sharing in a fair and equitable way the results of research and development and the benefits arising from the commercial and other utilization of genetic resources with the Contracting Party providing such resources. Such sharing shall be upon mutually agreed terms.

Article 16. Access to and Transfer of Technology

1. Each Contracting Party, recognizing that technology includes biotechnology, and that both access to and transfer of technology among Contracting Parties are essential elements for the attainment of the objectives of this Convention, undertakes subject to the provisions of this Article to provide and/or facilitate access for and transfer to other Contracting Parties of technologies that are relevant to the conservation and sustainable use of biological diversity or make use of genetic resources and do not cause significant damage to the environment.
2. Access to and transfer of technology referred to in paragraph 1 above to developing countries shall be provided and/or facilitated under fair and most favourable terms, including on concessional and preferential terms where mutually agreed, and, where necessary, in accordance with the financial mechanism established by Articles 20 and 21. In the case of technology subject to patents and other intellectual property rights, such access and transfer shall be provided on terms which recognize and are consistent with the adequate and effective protection of intellectual property rights. The application of this paragraph shall be consistent with paragraphs 3, 4 and 5 below.
3. Each Contracting Party shall take legislative, administrative or policy measures, as appropriate, with the aim that Contracting Parties, in particular those that are developing countries, which provide genetic resources are provided access to and transfer of technology which makes use of those resources, on mutually agreed terms, including technology protected by patents and other intellectual property rights, where necessary, through the provisions of Articles 20 and 21 and in accordance with international law and consistent with paragraphs 4 and 5 below.
4. Each Contracting Party shall take legislative, administrative or policy measures, as appropriate, with the aim that the private sector facilitates access to, joint development and transfer of technology referred to in

paragraph 1 above for the benefit of both governmental institutions and the private sector of developing countries and in this regard shall abide by the obligations included in paragraphs 1, 2 and 3 above.

5. The Contracting Parties, recognizing that patents and other intellectual property rights may have an influence on the implementation of this Convention, shall cooperate in this regard subject to national legislation and international law in order to ensure that such rights are supportive of and do not run counter to its objectives.

Article 17. Exchange of Information

1. The Contracting Parties shall facilitate the exchange of information, from all publicly available sources, relevant to the conservation and sustainable use of biological diversity, taking into account the special needs of developing countries.
2. Such exchange of information shall include exchange of results of technical, scientific and socio-economic research, as well as information on training and surveying programmes, specialized knowledge, indigenous and traditional knowledge as such and in combination with the technologies referred to in Article 16, paragraph 1. It shall also, where feasible, include repatriation of information.

Article 18. Technical and Scientific Cooperation

1. The Contracting Parties shall promote international technical and scientific cooperation in the field of conservation and sustainable use of biological diversity, where necessary, through the appropriate international and national institutions.
2. Each Contracting Party shall promote technical and scientific cooperation with other Contracting Parties, in particular developing countries, in implementing this Convention, inter alia, through the development and implementation of national policies. In promoting such cooperation, special attention should be given to the development and strengthening of national capabilities, by means of human resources development and institution building.
3. The Conference of the Parties, at its first meeting, shall determine how to establish a clearing-house mechanism to promote and facilitate technical and scientific cooperation.

4. The Contracting Parties shall, in accordance with national legislation and policies, encourage and develop methods of cooperation for the development and use of technologies, including indigenous and traditional technologies, in pursuance of the objectives of this Convention. For this purpose, the Contracting Parties shall also promote cooperation in the training of personnel and exchange of experts.
5. The Contracting Parties shall, subject to mutual agreement, promote the establishment of joint research programmes and joint ventures for the development of technologies relevant to the objectives of this Convention.

Article 19. Handling of Biotechnology and Distribution of its Benefits

1. Each Contracting Party shall take legislative, administrative or policy measures, as appropriate, to provide for the effective participation in biotechnological research activities by those Contracting Parties, especially developing countries, which provide the genetic resources for such research, and where feasible in such Contracting Parties.
2. Each Contracting Party shall take all practicable measures to promote and advance priority access on a fair and equitable basis by Contracting Parties, especially developing countries, to the results and benefits arising from biotechnologies based upon genetic resources provided by those Contracting Parties. Such access shall be on mutually agreed terms.
3. The Parties shall consider the need for and modalities of a protocol setting out appropriate procedures, including, in particular, advance informed agreement, in the field of the safe transfer, handling and use of any living modified organism resulting from biotechnology that may have adverse effect on the conservation and sustainable use of biological diversity.
4. Each Contracting Party shall, directly or by requiring any natural or legal person under its jurisdiction providing the organisms referred to in paragraph 3 above, provide any available information about the use and safety regulations required by that Contracting Party in handling such organisms, as well as any available information on the potential adverse impact of the specific organisms concerned to the Contracting Party into which those organisms are to be introduced.

Article 20. Financial Resources

1. Each Contracting Party undertakes to provide, in accordance with its capabilities, financial support and incentives in respect of those national activities which are intended to achieve the objectives of this Convention, in accordance with its national plans, priorities and programmes.
2. The developed country Parties shall provide new and additional financial resources to enable developing country Parties to meet the agreed full incremental costs to them of implementing measures which fulfil the obligations of this Convention and to benefit from its provisions and which costs are agreed between a developing country Party and the institutional structure referred to in Article 21, in accordance with policy, strategy, programme priorities and eligibility criteria and an indicative list of incremental costs established by the Conference of the Parties. Other Parties, including countries undergoing the process of transition to a market economy, may voluntarily assume the obligations of the developed country Parties. For the purpose of this Article, the Conference of the Parties, shall at its first meeting establish a list of developed country Parties and other Parties which voluntarily assume the obligations of the developed country Parties. The Conference of the Parties shall periodically review and if necessary amend the list. Contributions from other countries and sources on a voluntary basis would also be encouraged. The implementation of these commitments shall take into account the need for adequacy, predictability and timely flow of funds and the importance of burden-sharing among the contributing Parties included in the list.
3. The developed country Parties may also provide, and developing country Parties avail themselves of, financial resources related to the implementation of this Convention through bilateral, regional and other multilateral channels.
4. The extent to which developing country Parties will effectively implement their commitments under this Convention will depend on the effective implementation by developed country Parties of their commitments under this Convention related to financial resources and transfer of technology and will take fully into account the fact that economic and social development and eradication of poverty are the first and overriding priorities of the developing country Parties.

5. The Parties shall take full account of the specific needs and special situation of least developed countries in their actions with regard to funding and transfer of technology.
6. The Contracting Parties shall also take into consideration the special conditions resulting from the dependence on, distribution and location of, biological diversity within developing country Parties, in particular small island States.
7. Consideration shall also be given to the special situation of developing countries, including those that are most environmentally vulnerable, such as those with arid and semi-arid zones, coastal and mountainous areas.

Article 21. Financial Mechanism

1. There shall be a mechanism for the provision of financial resources to developing country Parties for purposes of this Convention on a grant or concessional basis the essential elements of which are described in this Article. The mechanism shall function under the authority and guidance of, and be accountable to, the Conference of the Parties for purposes of this Convention. The operations of the mechanism shall be carried out by such institutional structure as may be decided upon by the Conference of the Parties at its first meeting. For purposes of this Convention, the Conference of the Parties shall determine the policy, strategy, programme priorities and eligibility criteria relating to the access to and utilization of such resources. The contributions shall be such as to take into account the need for predictability, adequacy and timely flow of funds referred to in Article 20 in accordance with the amount of resources needed to be decided periodically by the Conference of the Parties and the importance of burden-sharing among the contributing Parties included in the list referred to in Article 20, paragraph 2. Voluntary contributions may also be made by the developed country Parties and by other countries and sources. The mechanism shall operate within a democratic and transparent system of governance.
2. Pursuant to the objectives of this Convention, the Conference of the Parties shall at its first meeting determine the policy, strategy and programme priorities, as well as detailed criteria and guidelines for eligibility for access to and utilization of the financial resources

including monitoring and evaluation on a regular basis of such utilization. The Conference of the Parties shall decide on the arrangements to give effect to paragraph 1 above after consultation with the institutional structure entrusted with the operation of the financial mechanism.

3. The Conference of the Parties shall review the effectiveness of the mechanism established under this Article, including the criteria and guidelines referred to in paragraph 2 above, not less than two years after the entry into force of this Convention and thereafter on a regular basis. Based on such review, it shall take appropriate action to improve the effectiveness of the mechanism if necessary.
4. The Contracting Parties shall consider strengthening existing financial institutions to provide financial resources for the conservation and sustainable use of biological diversity.

Article 22. Relationship with Other International Conventions

1. The provisions of this Convention shall not affect the rights and obligations of any Contracting Party deriving from any existing international agreement, except where the exercise of those rights and obligations would cause a serious damage or threat to biological diversity.
2. Contracting Parties shall implement this Convention with respect to the marine environment consistently with the rights and obligations of States under the law of the sea.

Article 23. Conference of the Parties

1. A Conference of the Parties is hereby established. The first meeting of the Conference of the Parties shall be convened by the Executive Director of the United Nations Environment Programme not later than one year after the entry into force of this Convention. Thereafter, ordinary meetings of the Conference of the Parties shall be held at regular intervals to be determined by the Conference at its first meeting.
2. Extraordinary meetings of the Conference of the Parties shall be held at such other times as may be deemed necessary by the Conference, or at the written request of any Party, provided that, within six months of the request being communicated to them by the Secretariat, it is supported by at least one third of the Parties.

3. The Conference of the Parties shall by consensus agree upon and adopt rules of procedure for itself and for any subsidiary body it may establish, as well as financial rules governing the funding of the Secretariat. At each ordinary meeting, it shall adopt a budget for the financial period until the next ordinary meeting.

4. The Conference of the Parties shall keep under review the implementation of this Convention, and, for this purpose, shall:

   (a) Establish the form and the intervals for transmitting the information to be submitted in accordance with Article 26 and consider such information as well as reports submitted by any subsidiary body;

   (b) Review scientific, technical and technological advice on biological diversity provided in accordance with Article 25;

   (c) Consider and adopt, as required, protocols in accordance with Article 28;

   (d) Consider and adopt, as required, in accordance with Articles 29 and 30, amendments to this Convention and its annexes;

   (e) Consider amendments to any protocol, as well as to any annexes thereto, and, if so decided, recommend their adoption to the parties to the protocol concerned;

   (f) Consider and adopt, as required, in accordance with Article 30, additional annexes to this Convention;

   (g) Establish such subsidiary bodies, particularly to provide scientific and technical advice, as are deemed necessary for the implementation of this Convention;

   (h) Contact, through the Secretariat, the executive bodies of conventions dealing with matters covered by this Convention with a view to establishing appropriate forms of cooperation with them; and

   (i) Consider and undertake any additional action that may be required for the achievement of the purposes of this Convention in the light of experience gained in its operation.

5. The United Nations, its specialized agencies and the International Atomic Energy Agency, as well as any State not Party to this Convention, may be represented as observers at meetings of the

Conference of the Parties. Any other body or agency, whether governmental or non-governmental, qualified in fields relating to conservation and sustainable use of biological diversity, which has informed the Secretariat of its wish to be represented as an observer at a meeting of the Conference of the Parties, may be admitted unless at least one third of the Parties present object. The admission and participation of observers shall be subject to the rules of procedure adopted by the Conference of the Parties.

Article 24. Secretariat

1. A secretariat is hereby established. Its functions shall be:
   (a) To arrange for and service meetings of the Conference of the Parties provided for in Article 23;
   (b) To perform the functions assigned to it by any protocol;
   (c) To prepare reports on the execution of its functions under this Convention and present them to the Conference of the Parties;
   (d) To coordinate with other relevant international bodies and, in particular to enter into such administrative and contractual arrangements as may be required for the effective discharge of its functions; and
   (e) To perform such other functions as may be determined by the Conference of the Parties.
2. At its first ordinary meeting, the Conference of the Parties shall designate the secretariat from amongst those existing competent international organizations which have signified their willingness to carry out the secretariat functions under this Convention.

Article 25. Subsidiary Body on Scientific, Technical and Technological Advice

1. A subsidiary body for the provision of scientific, technical and technological advice is hereby established to provide the Conference of the Parties and, as appropriate, its other subsidiary bodies with timely advice relating to the implementation of this Convention. This body shall be open to participation by all Parties and shall be multidisciplinary. It shall comprise government representatives competent in the relevant field of expertise. It shall report regularly to the Conference of the Parties on all aspects of its work.

2. Under the authority of and in accordance with guidelines laid down by the Conference of the Parties, and upon its request, this body shall:
   (a) Provide scientific and technical assessments of the status of biological diversity;
   (b) Prepare scientific and technical assessments of the effects of types of measures taken in accordance with the provisions of this Convention;
   (c) Identify innovative, efficient and state-of-the-art technologies and know-how relating to the conservation and sustainable use of biological diversity and advise on the ways and means of promoting development and/or transferring such technologies;
   (d) Provide advice on scientific programmes and international cooperation in research and development related to conservation and sustainable use of biological diversity; and
   (e) Respond to scientific, technical, technological and methodological questions that the Conference of the Parties and its subsidiary bodies may put to the body.
3. The functions, terms of reference, organization and operation of this body may be further elaborated by the Conference of the Parties.

Article 26. Reports

Each Contracting Party shall, at intervals to be determined by the Conference of the Parties, present to the Conference of the Parties, reports on measures which it has taken for the implementation of the provisions of this Convention and their effectiveness in meeting the objectives of this Convention.

Article 27. Settlement of Disputes

1. In the event of a dispute between Contracting Parties concerning the interpretation or application of this Convention, the parties concerned shall seek solution by negotiation.
2. If the parties concerned cannot reach agreement by negotiation, they may jointly seek the good offices of, or request mediation by, a third party.
3. When ratifying, accepting, approving or acceding to this Convention, or at any time thereafter, a State or regional economic integration

organization may declare in writing to the Depositary that for a dispute not resolved in accordance with paragraph 1 or paragraph 2 above, it accepts one or both of the following means of dispute settlement as compulsory:

(a) Arbitration in accordance with the procedure laid down in Part 1 of Annex II;

(b) Submission of the dispute to the International Court of Justice.

4. If the parties to the dispute have not, in accordance with paragraph 3 above, accepted the same or any procedure, the dispute shall be submitted to conciliation in accordance with Part 2 of Annex II unless the parties otherwise agree.

5. The provisions of this Article shall apply with respect to any protocol except as otherwise provided in the protocol concerned.

Article 28. Adoption of Protocols

1. The Contracting Parties shall cooperate in the formulation and adoption of protocols to this Convention.

2. Protocols shall be adopted at a meeting of the Conference of the Parties.

3. The text of any proposed protocol shall be communicated to the Contracting Parties by the Secretariat at least six months before such a meeting.

Article 29. Amendment of the Convention or Protocols

1. Amendments to this Convention may be proposed by any Contracting Party. Amendments to any protocol may be proposed by any Party to that protocol.

2. Amendments to this Convention shall be adopted at a meeting of the Conference of the Parties. Amendments to any protocol shall be adopted at a meeting of the Parties to the Protocol in question. The text of any proposed amendment to this Convention or to any protocol, except as may otherwise be provided in such protocol, shall be communicated to the Parties to the instrument in question by the secretariat at least six months before the meeting at which it is proposed for adoption. The secretariat shall also communicate proposed amendments to the signatories to this Convention for information.

3. The Parties shall make every effort to reach agreement on any proposed amendment to this Convention or to any protocol by consensus. If all efforts at consensus have been exhausted, and no agreement reached, the amendment shall as a last resort be adopted by a two-third majority vote of the Parties to the instrument in question present and voting at the meeting, and shall be submitted by the Depositary to all Parties for ratification, acceptance or approval.
4. Ratification, acceptance or approval of amendments shall be notified to the Depositary in writing. Amendments adopted in accordance with paragraph 3 above shall enter into force among Parties having accepted them on the ninetieth day after the deposit of instruments of ratification, acceptance or approval by at least two thirds of the Contracting Parties to this Convention or of the Parties to the protocol concerned, except as may otherwise be provided in such protocol. Thereafter the amendments shall enter into force for any other Party on the ninetieth day after that Party deposits its instrument of ratification, acceptance or approval of the amendments.
5. For the purposes of this Article, "Parties present and voting" means Parties present and casting an affirmative or negative vote.

Article 30. Adoption and Amendment of Annexes

1. The annexes to this Convention or to any protocol shall form an integral part of the Convention or of such protocol, as the case may be, and, unless expressly provided otherwise, a reference to this Convention or its protocols constitutes at the same time a reference to any annexes thereto. Such annexes shall be restricted to procedural, scientific, technical and administrative matters.
2. Except as may be otherwise provided in any protocol with respect to its annexes, the following procedure shall apply to the proposal, adoption and entry into force of additional annexes to this Convention or of annexes to any protocol:
   (a) Annexes to this Convention or to any protocol shall be proposed and adopted according to the procedure laid down in Article 29;
   (b) Any Party that is unable to approve an additional annex to this Convention or an annex to any protocol to which it is Party shall so notify the Depositary, in writing, within one year from the date

of the communication of the adoption by the Depositary. The Depositary shall without delay notify all Parties of any such notification received. A Party may at any time withdraw a previous declaration of objection and the annexes shall thereupon enter into force for that Party subject to subparagraph (c) below;

(c) On the expiry of one year from the date of the communication of the adoption by the Depositary, the annex shall enter into force for all Parties to this Convention or to any protocol concerned which have not submitted a notification in accordance with the provisions of subparagraph (b) above.

3. The proposal, adoption and entry into force of amendments to annexes to this Convention or to any protocol shall be subject to the same procedure as for the proposal, adoption and entry into force of annexes to the Convention or annexes to any protocol.

4. If an additional annex or an amendment to an annex is related to an amendment to this Convention or to any protocol, the additional annex or amendment shall not enter into force until such time as the amendment to the Convention or to the protocol concerned enters into force.

Article 31. Right to Vote

1. Except as provided for in paragraph 2 below, each Contracting Party to this Convention or to any protocol shall have one vote.

2. Regional economic integration organizations, in matters within their competence, shall exercise their right to vote with a number of votes equal to the number of their member States which are Contracting Parties to this Convention or the relevant protocol. Such organizations shall not exercise their right to vote if their member States exercise theirs, and vice versa.

Article 32. Relationship between this Convention and Its Protocols

1. A State or a regional economic integration organization may not become a Party to a protocol unless it is, or becomes at the same time, a Contracting Party to this Convention.

2. Decisions under any protocol shall be taken only by the Parties to the protocol concerned. Any Contracting Party that has not ratified,

accepted or approved a protocol may participate as an observer in any meeting of the parties to that protocol.

Article 33. Signature

This Convention shall be open for signature at Rio de Janeiro by all States and any regional economic integration organization from 5 June 1992 until 14 June 1992, and at the United Nations Headquarters in New York from 15 June 1992 to 4 June 1993.

Article 34. Ratification, Acceptance or Approval

1. This Convention and any protocol shall be subject to ratification, acceptance or approval by States and by regional economic integration organizations. Instruments of ratification, acceptance or approval shall be deposited with the Depositary.
2. Any organization referred to in paragraph 1 above which becomes a Contracting Party to this Convention or any protocol without any of its member States being a Contracting Party shall be bound by all the obligations under the Convention or the protocol, as the case may be. In the case of such organizations, one or more of whose member States is a Contracting Party to this Convention or relevant protocol, the organization and its member States shall decide on their respective responsibilities for the performance of their obligations under the Convention or protocol, as the case may be. In such cases, the organization and the member States shall not be entitled to exercise rights under the Convention or relevant protocol concurrently.
3. In their instruments of ratification, acceptance or approval, the organizations referred to in paragraph 1 above shall declare the extent of their competence with respect to the matters governed by the Convention or the relevant protocol. These organizations shall also inform the Depositary of any relevant modification in the extent of their competence.

Article 35. Accession

1. This Convention and any protocol shall be open for accession by States and by regional economic integration organizations from the date on which the Convention or the protocol concerned is closed for signature. The instruments of accession shall be deposited with the Depositary.

2. In their instruments of accession, the organizations referred to in paragraph 1 above shall declare the extent of their competence with respect to the matters governed by the Convention or the relevant protocol. These organizations shall also inform the Depositary of any relevant modification in the extent of their competence.
3. The provisions of Article 34, paragraph 2, shall apply to regional economic integration organizations which accede to this Convention or any protocol.

Article 36. Entry Into Force

1. This Convention shall enter into force on the ninetieth day after the date of deposit of the thirtieth instrument of ratification, acceptance, approval or accession.
2. Any protocol shall enter into force on the ninetieth day after the date of deposit of the number of instruments of ratification, acceptance, approval or accession, specified in that protocol, has been deposited.
3. For each Contracting Party which ratifies, accepts or approves this Convention or accedes thereto after the deposit of the thirtieth instrument of ratification, acceptance, approval or accession, it shall enter into force on the ninetieth day after the date of deposit by such Contracting Party of its instrument of ratification, acceptance, approval or accession.
4. Any protocol, except as otherwise provided in such protocol, shall enter into force for a Contracting Party that ratifies, accepts or approves that protocol or accedes thereto after its entry into force pursuant to paragraph 2 above, on the ninetieth day after the date on which that Contracting Party deposits its instrument of ratification, acceptance, approval or accession, or on the date on which this Convention enters into force for that Contracting Party, whichever shall be the later.
5. For the purposes of paragraphs 1 and 2 above, any instrument deposited by a regional economic integration organization shall not be counted as additional to those deposited by member States of such organization.

Article 37. Reservations

No reservations may be made to this Convention.

Article 38. Withdrawals

1. At any time after two years from the date on which this Convention has entered into force for a Contracting Party, that Contracting Party may withdraw from the Convention by giving written notification to the Depositary.
2. Any such withdrawal shall take place upon expiry of one year after the date of its receipt by the Depositary, or on such later date as may be specified in the notification of the withdrawal.
3. Any Contracting Party which withdraws from this Convention shall be considered as also having withdrawn from any protocol to which it is party.

Article 39. Financial Interim Arrangements

Provided that it has been fully restructured in accordance with the requirements of Article 21, the Global Environment Facility of the United Nations Development Programme, the United Nations Environment Programme and the International Bank for Reconstruction and Development shall be the institutional structure referred to in Article 21 on an interim basis, for the period between the entry into force of this Convention and the first meeting of the Conference of the Parties or until the Conference of the Parties decides which institutional structure will be designated in accordance with Article 21.

Article 40. Secretariat Interim Arrangements

The secretariat to be provided by the Executive Director of the United Nations Environment Programme shall be the secretariat referred to in Article 24, paragraph 2, on an interim basis for the period between the entry into force of this Convention and the first meeting of the Conference of the Parties.

Article 41. Depositary

The Secretary-General of the United Nations shall assume the functions of Depositary of this Convention and any protocols.

Article 42. Authentic Texts

The original of this Convention, of which the Arabic, Chinese, English,

French, Russian and Spanish texts are equally authentic, shall be deposited with the Secretary-General of the United Nations. In witness whereof the undersigned, being duly authorized to that effect, have signed this Convention.

Done at Rio de Janeiro on this fifth day of June, one thousand nine hundred and ninety-two.

Annex I Identification and Monitoring

1. Ecosystems and habitats: containing high diversity, large numbers of endemic or threatened species, or wilderness; required by migratory species; of social, economic, cultural or scientific importance; or, which are representative, unique or associated with key evolutionary or other biological processes;
2. Species and communities which are: threatened; wild relatives of domesticated or cultivated species; of medicinal, agricultural or other economic value; or social, scientific or cultural importance; or importance for research into the conservation and sustainable use of biological diversity, such as indicator species; and
3. Described genomes and genes of social, scientific or economic importance.

Annex II
Part 1 Arbitration

Article 1

The claimant party shall notify the secretariat that the parties are referring a dispute to arbitration pursuant to Article 27. The notification shall state the subject-matter of arbitration and include, in particular, the articles of the Convention or the protocol, the interpretation or application of which are at issue. If the parties do not agree on the subject matter of the dispute before the President of the tribunal is designated, the arbitral tribunal shall determine the subject matter. The secretariat shall forward the information thus received to all Contracting Parties to this Convention or to the protocol concerned.

Article 2

1. In disputes between two parties, the arbitral tribunal shall consist of three members. Each of the parties to the dispute shall appoint an

arbitrator and the two arbitrators so appointed shall designate by common agreement the third arbitrator who shall be the President of the tribunal. The latter shall not be a national of one of the parties to the dispute, nor have his or her usual place of residence in the territory of one of these parties, nor be employed by any of them, nor have dealt with the case in any other capacity.

2. In disputes between more than two parties, parties in the same interest shall appoint one arbitrator jointly by agreement.
3. Any vacancy shall be filled in the manner prescribed for the initial appointment.

Article 3

1. If the President of the arbitral tribunal has not been designated within two months of the appointment of the second arbitrator, the Secretary-General of the United Nations shall, at the request of a party, designate the President within a further two-month period.
2. If one of the parties to the dispute does not appoint an arbitrator within two months of receipt of the request, the other party may inform the Secretary-General who shall make the designation within a further two-month period.

Article 4

The arbitral tribunal shall render its decisions in accordance with the provisions of this Convention, any protocols concerned, and international law.

Article 5

Unless the parties to the dispute otherwise agree, the arbitral tribunal shall determine its own rules of procedure.

Article 6

The arbitral tribunal may, at the request of one of the parties, recommend essential interim measures of protection.

Article 7

The parties to the dispute shall facilitate the work of the arbitral tribunal and, in particular, using all means at their disposal, shall:

(a) Provide it with all relevant documents, information and facilities; and

(b) Enable it, when necessary, to call witnesses or experts and receive their evidence.

Article 8

The parties and the arbitrators are under an obligation to protect the confidentiality of any information they receive in confidence during the proceedings of the arbitral tribunal.

Article 9

Unless the arbitral tribunal determines otherwise because of the particular circumstances of the case, the costs of the tribunal shall be borne by the parties to the dispute in equal shares. The tribunal shall keep a record of all its costs, and shall furnish a final statement thereof to the parties.

Article 10

Any Contracting Party that has an interest of a legal nature in the subject-matter of the dispute which may be affected by the decision in the case, may intervene in the proceedings with the consent of the tribunal.

Article 11

The tribunal may hear and determine counterclaims arising directly out of the subject-matter of the dispute.

Article 12

Decisions both on procedure and substance of the arbitral tribunal shall be taken by a majority vote of its members.

Article 13

If one of the parties to the dispute does not appear before the arbitral tribunal or fails to defend its case, the other party may request the tribunal to continue the proceedings and to make its award. Absence of a party or a failure of a party to defend its case shall not constitute a bar to the proceedings. Before rendering its final decision, the arbitral tribunal must satisfy itself that the claim is well founded in fact and law.

Article 14

The tribunal shall render its final decision within five months of the date on which it is fully constituted unless it finds it necessary to extend the time-limit for a period which should not exceed five more months.

Article 15

The final decision of the arbitral tribunal shall be confined to the subject-matter of the dispute and shall state the reasons on which it is based. It shall contain the names of the members who have participated and the date of the final decision. Any member of the tribunal may attach a separate or dissenting opinion to the final decision.

Article 16

The award shall be binding on the parties to the dispute. It shall be without appeal unless the parties to the dispute have agreed in advance to an appellate procedure.

Article 17

Any controversy which may arise between the parties to the dispute as regards the interpretation or manner of implementation of the final decision may be submitted by either party for decision to the arbitral tribunal which rendered it.

Part 2
Conciliation

Article 1

A conciliation commission shall be created upon the request of one of the parties to the dispute. The commission shall, unless the parties otherwise agree, be composed of five members, two appointed by each Party concerned and a President chosen jointly by those members.

Article 2

In disputes between more than two parties, parties in the same interest shall appoint their members of the commission jointly by agreement. Where two or more parties have separate interests or there is a disagreement as to whether they are of the same interest, they shall appoint their members separately.

Article 3

If any appointments by the parties are not made within two months of the date of the request to create a conciliation commission, the Secretary-General of the United Nations shall, if asked to do so by the party that made the request, make those appointments within a further two-month period.

Article 4

If a President of the conciliation commission has not been chosen within two months of the last of the members of the commission being appointed, the Secretary-General of the United Nations shall, if asked to do so by a party, designate a President within a further two-month period.

Article 5

The conciliation commission shall take its decisions by majority vote of its members. It shall, unless the parties to the dispute otherwise agree, determine its own procedure. It shall render a proposal for resolution of the dispute, which the parties shall consider in good faith.

Article 6

A disagreement as to whether the conciliation commission has competence shall be decided by the commission.

9

Cartagena Protocol on Biosafety to the Convention on Biological Diversity

The Cartagena Protocol on Biosafety is an international agreement on biosafety, as a supplement to the Convention on Biological Diversity. The Biosafety Protocol seeks to protect biological diversity from the potential risks posed by genetically modified organisms resulting from modern biotechnology.

The Biosafety Protocol makes clear that products from new technologies must be based on the precautionary principle and allow developing nations to balance public health against economic benefits. It will for example let countries ban imports of a genetically modified organisms if they feel there is not enough scientific evidence that the product is safe and requires exporters to label shipments containing genetically altered commodities such as corn or cotton.

The required number of 50 instruments of ratification/accession/approval/acceptance by countries was reached in May 2003. In accordance with the provisions of its Article 37, the Protocol entered into force on 11 September 2003.

In accordance with the precautionary approach, contained in Principle 15 of the Rio Declaration on Environment and Development, the objective of the Protocol is to contribute to ensuring an adequate level of protection in the field of the safe transfer, handling and use of 'living modified organisms resulting from modern biotechnology' that may have adverse effects on the conservation and sustainable use of biological diversity, taking also into account risks to human health, and specifically focusing on transboundary movements.

The fulltext of the Protocol is give below.

PREAMBLE

The Parties to this Protocol,

Being Parties to the Convention on Biological Diversity, hereinafter referred to as "the Convention",

Recalling Article 19, paragraphs 3 and 4, and Articles 8(g) and 17 of the Convention,

Recalling also decision II/5 of 17 November 1995 of the Conference of the Parties to the Convention to develop a Protocol on biosafety, specifically focusing on trans-boundary movement of any living modified organism resulting from modern biotechnology that may have adverse effect on the conservation and sustainable use of biological diversity, setting out for consideration, in particular, appropriate procedures for advance informed agreement,

Reaffirming the precautionary approach contained in Principle 15 of the Rio Declaration on Environment and Development,

Aware of the rapid expansion of modern biotechnology and the growing public concern over its potential adverse effects on biological diversity, taking also into account risks to human health,

Recognizing that modern biotechnology has great potential for human well-being if developed and used with adequate safety measures for the environment and human health,

Recognizing also the crucial importance to humankind of centres of origin and centres of genetic diversity,

Taking into account the limited capabilities of many countries, particularly developing countries, to cope with the nature and scale of known and potential risks associated with living modified organisms,

Recognizing that trade and environment agreements should be mutually supportive with a view to achieving sustainable development,

Emphasizing that this Protocol shall not be interpreted as implying a change in the rights and obligations of a Party under any existing international agreements,

Understanding that the above recital is not intended to subordinate this Protocol to other international agreements,

Have agreed as follows:

Article 1. Objective

In accordance with the precautionary approach contained in Principle 15 of the Rio Declaration on Environment and Development, the objective of this Protocol is to contribute to ensuring an adequate level of protection in the field of the safe transfer, handling and use of living modified organisms resulting from modern biotechnology that may have adverse effects on the conservation and sustainable use of biological diversity, taking also into account risks to human health, and specifically focusing on transboundary movements.

Article 2. General Provisions

1. Each Party shall take necessary and appropriate legal, administrative and other measures to implement its obligations under this Protocol.
2. The Parties shall ensure that the development, handling, transport, use, transfer and release of any living modified organisms are undertaken in a manner that prevents or reduces the risks to biological diversity, taking also into account risks to human health.
3. Nothing in this Protocol shall affect in any way the sovereignty of States over their territorial sea established in accordance with international law, and the sovereign rights and the jurisdiction which States have in their exclusive economic zones and their continental shelves in accordance with international law, and the exercise by ships and aircraft of all States of navigational rights and freedoms as provided for in international law and as reflected in relevant international instruments.
4. Nothing in this Protocol shall be interpreted as restricting the right of a Party to take action that is more protective of the conservation and sustainable use of biological diversity than that called for in this Protocol, provided that such action is consistent with the objective and the provisions of this Protocol and is in accordance with that Party's other obligations under international law.
5. The Parties are encouraged to take into account, as appropriate, available expertise, instruments and work undertaken in international forums with competence in the area of risks to human health.

ARTICLE 3. USE OF TERMS

For the purposes of this Protocol:

(a) "Conference of the Parties" means the Conference of the Parties to the Convention;

(b) "Contained use" means any operation, undertaken within a facility, installation or other physical structure, which involves living modified organisms that are controlled by specific measures that effectively limit their contact with, and their impact on, the external environment;

(c) "Export" means intentional transboundary movement from one Party to another Party;

(d) "Exporter" means any legal or natural person, under the jurisdiction of the Party of export, who arranges for a living modified organism to be exported;

(e) "Import" means intentional transboundary movement into one Party from another Party;

(f) "Importer" means any legal or natural person, under the jurisdiction of the Party of import, who arranges for a living modified organism to be imported;

(g) "Living modified organism" means any living organism that possesses a novel combination of genetic material obtained through the use of modern biotechnology;

(h) "Living organism" means any biological entity capable of transferring or replicating genetic material, including sterile organisms, viruses and viroids;

(i) "Modern biotechnology" means the application of:

(a) *In vitro* nucleic acid techniques, including recombinant deoxyribonucleic acid (DNA) and direct injection of nucleic acid into cells or organelles, or

(b) Fusion of cells beyond the taxonomic family, that overcome natural physiological reproductive or recombination barriers and that are not techniques used in traditional breeding and selection;

(j) "Regional economic integration organization" means an organization constituted by sovereign States of a given region, to which its member States have transferred competence in respect of matters governed by this Protocol and which has been duly authorized, in accordance with its internal procedures, to sign, ratify, accept, approve or accede to it;

(k) "Transboundary movement" means the movement of a living modified organism from one Party to another Party, save that for the purposes of Articles 17 and 24 transboundary movement extends to movement between Parties and non-Parties.

Article 4 . Scope

This Protocol shall apply to the transboundary movement, transit, handling and use of all living modified organisms that may have adverse effects on the conservation and sustainable use of biological diversity, taking also into account risks to human health.

Article 5. Pharmaceuticals

Notwithstanding Article 4 and without prejudice to any right of a Party to subject all living modified organisms to risk assessment prior to the making of decisions on import, this Protocol shall not apply to the transboundary movement of living modified organisms which are pharmaceuticals for humans that are addressed by other relevant international agreements or organisations.

Article 6. Transit and Contained Use

1. Notwithstanding Article 4 and without prejudice to any right of a Party of transit to regulate the transport of living modified organisms through its territory and make available to the Biosafety Clearing-House, any decision of that Party, subject to Article 2, paragraph 3, regarding the transit through its territory of a specific living modified organism, the provisions of this Protocol with respect to the advance informed agreement procedure shall not apply to living modified organisms in transit.
2. Notwithstanding Article 4 and without prejudice to any right of a Party to subject all living modified organisms to risk assessment prior to decisions on import and to set standards for contained use within its jurisdiction, the provisions of this Protocol with respect to the advance informed agreement procedure shall not apply to the transboundary movement of living modified organisms destined for contained use undertaken in accordance with the standards of the Party of import.

Article 7. Application of the Advance Informed Agreement Procedure

1. Subject to Articles 5 and 6, the advance informed agreement procedure in Articles 8 to 10 and 12 shall apply prior to the first intentional transboundary movement of living modified organisms for intentional introduction into the environment of the Party of import.
2. "Intentional introduction into the environment" in paragraph 1 above, does not refer to living modified organisms intended for direct use as food or feed, or for processing.
3. Article 11 shall apply prior to the first transboundary movement of living modified organisms intended for direct use as food or feed, or for processing.
4. The advance informed agreement procedure shall not apply to the intentional transboundary movement of living modified organisms identified in a decision of the Conference of the Parties serving as the meeting of the Parties to this Protocol as being not likely to have adverse effects on the conservation and sustainable use of biological diversity, taking also into account risks to human health.

Article 8. Notification

1. The Party of export shall notify, or require the exporter to ensure notification to, in writing, the competent national authority of the Party of import prior to the intentional transboundary movement of a living modified organism that falls within the scope of Article 7, paragraph 1. The notification shall contain, at a minimum, the information specified in Annex I.
2. The Party of export shall ensure that there is a legal requirement for the accuracy of information provided by the exporter.

Article 9. Acknowledgement of Receipt of Notification

1. The Party of import shall acknowledge receipt of the notification, in writing, to the notifier within ninety days of its receipt.
2. The acknowledgement shall state:
   (a) The date of receipt of the notification;
   (b) Whether the notification, *prima facie,* contains the information referred to in Article 8;

(c) Whether to proceed according to the domestic regulatory framework of the Party of import or according to the procedure specified in Article 10.

3. The domestic regulatory framework referred to in paragraph 2(c) above, shall be consistent with this Protocol.

4. A failure by the Party of import to acknowledge receipt of a notification shall not imply its consent to an intentional transboundary movement.

Article 10. Decision Procedure

1. Decisions taken by the Party of import shall be in accordance with Article 15.

2. The Party of import shall, within the period of time referred to in Article 9, inform the notifier, in writing, whether the intentional transboundary movement may proceed:

   (a) Only after the Party of import has given its written consent; or

   (b) After no less than ninety days without a subsequent written consent.

3. Within two hundred and seventy days of the date of receipt of notification, the Party of import shall communicate, in writing, to the notifier and to the Biosafety Clearing-House the decision referred to in paragraph 2(a) above:

   (a) Approving the import, with or without conditions, including how the decision will apply to subsequent imports of the same living modified organism;

   (b) Prohibiting the import;

   (c) Requesting additional relevant information in accordance with its domestic regulatory framework or Annex I; in calculating the time within which the Party of import is to respond, the number of days it has to wait for additional relevant information shall not be taken into account; or

   (d) Informing the notifier that the period specified in this paragraph is extended by a defined period of time.

4. Except in a case in which consent is unconditional, a decision under paragraph 3 above, shall set out the reasons on which it is based.

5. A failure by the Party of import to communicate its decision within two hundred and seventy days of the date of receipt of the notification shall not imply its consent to an intentional transboundary movement.
6. Lack of scientific certainty due to insufficient relevant scientific information and knowledge regarding the extent of the potential adverse effects of a living modified organism on the conservation and sustainable use of biological diversity in the Party of import, taking also into account risks to human health, shall not prevent that Party from taking a decision, as appropriate, with regard to the import of the living modified organism in question as referred to in paragraph 3 above, in order to avoid or minimize such potential adverse effects.
7. The Conference of the Parties serving as the meeting of the Parties shall, at its first meeting, decide upon appropriate procedures and mechanisms to facilitate decision-making by Parties of import.

Article 11. Procedure for Living Modified Organisms Intended for Direct Use as Food or Feed, or for Processing

1. A Party that makes a final decision regarding domestic use, including placing on the market, of a living modified organism that may be subject to transboundary movement for direct use as food or feed, or for processing shall, within fifteen days of making that decision, inform the Parties through the Biosafety Clearing-House. This information shall contain, at a minimum, the information specified in Annex II. The Party shall provide a copy of the information, in writing, to the national focal point of each Party that informs the Secretariat in advance that it does not have access to the Biosafety Clearing-House. This provision shall not apply to decisions regarding field trials.
2. The Party making a decision under paragraph 1 above, shall ensure that there is a legal requirement for the accuracy of information provided by the applicant.
3. Any Party may request additional information from the authority identified in paragraph (b) of Annex II.
4. A Party may take a decision on the import of living modified organisms intended for direct use as food or feed, or for processing, under its domestic regulatory framework that is consistent with the objective of this Protocol.

5. Each Party shall make available to the Biosafety Clearing-House copies of any national laws, regulations and guidelines applicable to the import of living modified organisms intended for direct use as food or feed, or for processing, if available.
6. A developing country Party or a Party with an economy in transition may, in the absence of the domestic regulatory framework referred to in paragraph 4 above, and in exercise of its domestic jurisdiction, declare through the Biosafety Clearing-House that its decision prior to the first import of a living modified organism intended for direct use as food or feed, or for processing, on which information has been provided under paragraph 1 above, will be taken according to the following:
   (a) A risk assessment undertaken in accordance with Annex III; and
   (b) A decision made within a predictable timeframe, not exceeding two hundred and seventy days.
7. Failure by a Party to communicate its decision according to paragraph 6 above, shall not imply its consent or refusal to the import of a living modified organism intended for direct use as food or feed, or for processing, unless otherwise specified by the Party.
8. Lack of scientific certainty due to insufficient relevant scientific information and knowledge regarding the extent of the potential adverse effects of a living modified organism on the conservation and sustainable use of biological diversity in the Party of import, taking also into account risks to human health, shall not prevent that Party from taking a decision, as appropriate, with regard to the import of that living modified organism intended for direct use as food or feed, or for processing, in order to avoid or minimize such potential adverse effects.
9. A Party may indicate its needs for financial and technical assistance and capacity-building with respect to living modified organisms intended for direct use as food or feed, or for processing. Parties shall cooperate to meet these needs in accordance with Articles 22 and 28.

Article 12. Review of Decisions

1. A Party of import may, at any time, in light of new scientific information on potential adverse effects on the conservation and

sustainable use of biological diversity, taking also into account the risks to human health, review and change a decision regarding an intentional transboundary movement. In such case, the Party shall, within thirty days, inform any notifier that has previously notified movements of the living modified organism referred to in such decision, as well as the Biosafety Clearing-House, and shall set out the reasons for its decision.

2. A Party of export or a notifier may request the Party of import to review a decision it has made in respect of it under Article 10 where the Party of export or the notifier considers that:

   (a) A change in circumstances has occurred that may influence the outcome of the risk assessment upon which the decision was based; or

   (b) Additional relevant scientific or technical information has become available.

3. The Party of import shall respond in writing to such a request within ninety days and set out the reasons for its decision.

4. The Party of import may, at its discretion, require a risk assessment for subsequent imports.

Article 13. Simplified Procedure

1. A Party of import may, provided that adequate measures are applied to ensure the safe intentional transboundary movement of living modified organisms in accordance with the objective of this Protocol, specify in advance to the Biosafety Clearing-House:

   (a) Cases in which intentional transboundary movement to it may take place at the same time as the movement is notified to the Party of import; and

   (b) Imports of living modified organisms to it to be exempted from the advance informed agreement procedure. Notifications under subparagraph (a) above, may apply to subsequent similar movements to the same Party.

2. The information relating to an intentional transboundary movement that is to be provided in the notifications referred to in paragraph 1(a) above, shall be the information specified in Annex I.

Article 14. Bilateral, Regional and Multilateral Agreements and Arrangements

1. Parties may enter into bilateral, regional and multilateral agreements and arrangements regarding intentional transboundary movements of living modified organisms, consistent with the objective of this Protocol and provided that such agreements and arrangements do not result in a lower level of protection than that provided for by the Protocol.
2. The Parties shall inform each other, through the Biosafety Clearing-House, of any such bilateral, regional and multilateral agreements and arrangements that they have entered into before or after the date of entry into force of this Protocol.
3. The provisions of this Protocol shall not affect intentional transboundary movements that take place pursuant to such agreements and arrangements as between the parties to those agreements or arrangements.
4. Any Party may determine that its domestic regulations shall apply with respect to specific imports to it and shall notify the Biosafety Clearing-House of its decision.

Article 15. Risk Assessment

1. Risk assessments undertaken pursuant to this Protocol shall be carried out in a scientifically sound manner, in accordance with Annex III and taking into account recognized risk assessment techniques. Such risk assessments shall be based, at a minimum, on information provided in accordance with Article 8 and other available scientific evidence in order to identify and evaluate the possible adverse effects of living modified organisms on the conservation and sustainable use of biological diversity, taking also into account risks to human health.
2. The Party of import shall ensure that risk assessments are carried out for decisions taken under Article 10. It may require the exporter to carry out the risk assessment.
3. The cost of risk assessment shall be borne by the notifier if the Party of import so requires.

Article 16. Risk Management

1. The Parties shall, taking into account Article 8(g) of the Convention, establish and maintain appropriate mechanisms, measures and strategies

to regulate, manage and control risks identified in the risk assessment provisions of this Protocol associated with the use, handling and transboundary movement of living modified organisms.

2. Measures based on risk assessment shall be imposed to the extent necessary to prevent adverse effects of the living modified organism on the conservation and sustainable use of biological diversity, taking also into account risks to human health, within the territory of the Party of import.
3. Each Party shall take appropriate measures to prevent unintentional trans-boundary movements of living modified organisms, including such measures as requiring a risk assessment to be carried out prior to the first release of a living modified organism.
4. Without prejudice to paragraph 2 above, each Party shall endeavour to ensure that any living modified organism, whether imported or locally developed, has undergone an appropriate period of observation that is commensurate with its life-cycle or generation time before it is put to its intended use.
5. Parties shall cooperate with a view to:
   (a) Identifying living modified organisms or specific traits of living modified organisms that may have adverse effects on the conservation and sustainable use of biological diversity, taking also into account risks to human health; and
   (b) Taking appropriate measures regarding the treatment of such living modified organisms or specific traits.

Article 17. Unintentional Transboundary Movements and Emergency Measures

1. Each Party shall take appropriate measures to notify affected or potentially affected States, the Biosafety Clearing-House and, where appropriate, relevant international organizations, when it knows of an occurrence under its jurisdiction resulting in a release that leads, or may lead, to an unintentional transboundary movement of a living modified organism that is likely to have significant adverse effects on the conservation and sustainable use of biological diversity, taking also into account risks to human health in such States. The notification shall be provided as soon as the Party knows of the above situation.

2. Each Party shall, no later than the date of entry into force of this Protocol for it, make available to the Biosafety Clearing-House the relevant details setting out its point of contact for the purposes of receiving notifications under this Article.
3. Any notification arising from paragraph 1 above, should include:
   (a) Available relevant information on the estimated quantities and relevant characteristics and/or traits of the living modified organism;
   (b) Information on the circumstances and estimated date of the release, and on the use of the living modified organism in the originating Party;
   (c) Any available information about the possible adverse effects on the conservation and sustainable use of biological diversity, taking also into account risks to human health, as well as available information about possible risk management measures;
   (d) Any other relevant information; and
   (e) A point of contact for further information.
4. In order to minimize any significant adverse effects on the conservation and sustainable use of biological diversity, taking also into account risks to human health, each Party, under whose jurisdiction the release of the living modified organism referred to in paragraph 1 above, occurs, shall immediately consult the affected or potentially affected States to enable them to determine appropriateresponses and initiate necessary action, including emergency measures.

Article 18. Handling, Transport, Packaging and Identification

1. In order to avoid adverse effects on the conservation and sustainable use of biological diversity, taking also into account risks to human health, each Party shall take necessary measures to require that living modified organisms that are subject to intentional transboundary movement within the scope of this Protocol are handled, packaged and transported under conditions of safety, taking into consideration relevant international rules and standards.
2. Each Party shall take measures to require that documentation accompanying:

(a) Living modified organisms that are intended for direct use as food or feed, or for processing, clearly identifies that they "may contain" living modified organisms and are not intended for intentional introduction into the environment, as well as a contact point for further information. The Conference of the Parties serving as the meeting of the Parties to this Protocol shall take a decision on the detailed requirements for this purpose, including specification of their identity and any unique identification, no later than two years after the date of entry into force of this Protocol;

(b) Living modified organisms that are destined for contained use clearly identifies them as living modified organisms; and specifies any requirements for the safe handling, storage, transport and use, the contact point for further information, including the name and address of the individual and institution to whom the living modified organisms are consigned; and

(c) Living modified organisms that are intended for intentional introduction into the environment of the Party of import and any other living modified organisms within the scope of the Protocol, clearly identifies them as living modified organisms; specifies the identity and relevant traits and/or characteristics, any requirements for the safe handling, storage, transport and use, the contact point for further information and, as appropriate, the name and address of the importer and exporter; and contains a declaration that the movement is in conformity with the requirements of this Protocol applicable to the exporter.

3. The Conference of the Parties serving as the meeting of the Parties to this Protocol shall consider the need for and modalities of developing standards with regard to identification, handling, packaging and transport practices, in consultation with other relevant international bodies.

Article 19. Competent National Authorities and National Focal Points

1. Each Party shall designate one national focal point to be responsible on its behalf for liaison with the Secretariat. Each Party shall also designate one or more competent national authorities, which shall be

responsible for performing the administrative functions required by this Protocol and which shall be authorized to act on its behalf with respect to those functions. A Party may designate a single entity to fulfil the functions of both focal point and competent national authority.

2. Each Party shall, no later than the date of entry into force of this Protocol for it, notify the Secretariat of the names and addresses of its focal point and its competent national authority or authorities. Where a Party designates more than one competent national authority, it shall convey to the Secretariat, with its notification thereof, relevant information on the respective responsibilities of those authorities. Where applicable, such information shall, at a minimum, specify which competent authority is responsible for which type of living modified organism. Each Party shall forthwith notify the Secretariat of any changes in the designation of its national focal point or in the name and address or responsibilities of its competent national authority or authorities.

3. The Secretariat shall forthwith inform the Parties of the notifications it receives under paragraph 2 above, and shall also make such information available through the Biosafety Clearing-House.

Article 20. Information Sharing and the Biosafety Clearing-House

1. A Biosafety Clearing-House is hereby established as part of the clearing-house mechanism under Article 18, paragraph 3, of the Convention, in order to:

    (a) Facilitate the exchange of scientific, technical, environmental and legal information on, and experience with, living modified organisms; and

    (b) Assist Parties to implement the Protocol, taking into account the special needs of developing country Parties, in particular the least developed and small island developing States among them, and countries with economies in transition as well as countries that are centres of origin and centres of genetic diversity.

2. The Biosafety Clearing-House shall serve as a means through which information is made available for the purposes of paragraph 1 above. It shall provide access to information made available by the Parties

relevant to the implementation of the Protocol. It shall also provide access, where possible, to other international biosafety information exchange mechanisms.

3. Without prejudice to the protection of confidential information, each Party shall make available to the Biosafety Clearing-House any information required to be made available to the Biosafety Clearing-House under this Protocol, and:

   (a) Any existing laws, regulations and guidelines for implementation of the Protocol, as well as information required by the Parties for the advance informed agreement procedure;

   (b) Any bilateral, regional and multilateral agreements and arrangements;

   (c) Summaries of its risk assessments or environmental reviews of living modified organisms generated by its regulatory process, and carried out in accordance with Article 15, including, where appropriate, relevant information regarding products thereof, namely, processed materials that are of living modified organism origin, containing detectable novel combinations of replicable genetic material obtained through the use of modern biotechnology;

   (d) Its final decisions regarding the importation or release of living modified organisms; and

   (e) Reports submitted by it pursuant to Article *33,* including those on implementation of the advance informed agreement procedure.

4. The modalities of the operation of the Biosafety Clearing-House, including reports on its activities, shall be considered and decided upon by the Conference of the Parties serving as the meeting of the Parties to this Protocol at its first meeting, and kept under review thereafter.

Article 21. Confidential Information

1. The Party of import shall permit the notifier to identify information submitted under the procedures of this Protocol or required by the Party of import as part of the advance informed agreement procedure of the Protocol that is to be treated as confidential. Justification shall be given in such cases upon request.

2. The Party of import shall consult the notifier if it decides that information identified by the notifier as confidential does not qualify for such treatment and shall, prior to any disclosure, inform the notifier of its decision, providing reasons on request, as well as an opportunity for consultation and for an internal review of the decision prior to disclosure.
3. Each Party shall protect confidential information received under this Protocol, including any confidential information received in the context of the advance informed agreement procedure of the Protocol. Each Party shall ensure that it has procedures to protect such information and shall protect the confidentiality of such information in a manner no less favourable than its treatment of confidential information in connection with domestically produced living modified organisms.
4. The Party of import shall not use such information for a commercial purpose, except with the written consent of the notifier.
5. If a notifier withdraws or has withdrawn a notification, the Party of import shall respect the confidentiality of commercial and industrial information, including research and development information as well as information on which the Party and the notifier disagree as to its confidentiality.
6. Without prejudice to paragraph 5 above, the following information shall not be considered confidential:
   (a) The name and address of the notifier;
   (b) A general description of the living modified organism or organisms;
   (c) A summary of the risk assessment of the effects on the conservation and sustainable use of biological diversity, taking also into account risks to human health; and
   (d) Any methods and plans for emergency response.

Article 22. Capacity-Building

1. The Parties shall cooperate in the development and/or strengthening of human resources and institutional capacities in biosafety, including biotechnology to the extent that it is required for biosafety, for the purpose of the effective implementation of this Protocol, in developing country Parties, in particular the least developed and small island

developing States among them, and in Parties with economies in transition, including through existing global, regional, subregional and national institutions and organizations and, as appropriate, through facilitating private sector involvement.

2. For the purposes of implementing paragraph 1 above, in relation to cooperation, the needs of developing country Parties, in particular the least developed and small island developing States among them, for financial resources and access to and transfer of technology and know-how in accordance with the relevant provisions of the Convention, shall be taken fully into account for capacity-building in biosafety Cooperation in capacity-building shall, subject to the different situation, capabilities and requirements of each Party, include scientific and technical training in the proper and safe management of biotechnology, and in the use of risk assessment and risk management for biosafety, and the enhancement of technological and institutional capacities in biosafety. The needs of Parties with economies in transition shall also be taken fully into account for such capacity-building in biosafety.

Article 23. Public Awareness and Participation

1. The Parties shall:
   (a) Promote and facilitate public awareness, education and participation concerning the safe transfer, handling and use of living modified organisms in relation to the conservation and sustainable use of biological diversity, taking also into account risks to human health. In doing so, the Parties shall cooperate, as appropriate, with other States and international bodies;
   (b) Endeavour to ensure that public awareness and education encompass access to information on living modified organisms identified in accordance with this Protocol that may be imported.

2. The Parties shall, in accordance with their respective laws and regulations, consult the public in the decision-making process regarding living modified organisms and shall make the results of such decisions available to the public, while respecting confidential information in accordance with Article 21.

3. Each Party shall endeavour to inform its public about the means of public access to the Biosafety Clearing-House.

Article 24. Non-Parties

1. Transboundary movements of living modified organisms between Parties and non-Parties shall be consistent with the objective of this Protocol. The Parties may enter into bilateral, regional and multilateral agreements and arrangements with non-Parties regarding such transboundary movements.
2. The Parties shall encourage non-Parties to adhere to this Protocol and to contribute appropriate information to the Biosafety Clearing-House on living modified organisms released in, or moved into or out of, areas within their national jurisdictions.

Article 25. Illegal Transboundary Movements

1. Each Party shall adopt appropriate domestic measures aimed at preventing and, if appropriate, penalizing transboundary movements of living modified organisms carried out in contravention of its domestic measures to implement this Protocol. Such movements shall be deemed illegal transboundary movements.
2. In the case of an illegal transboundary movement, the affected Party may request the Party of origin to dispose, at its own expense, of the living modified organism in question by repatriation or destruction, as appropriate.
3. Each Party shall make available to the Biosafety Clearing-House information concerning cases of illegal transboundary movements pertaining to it.

Article 26. Socio-Economic Considerations

1. The Parties, in reaching a decision on import under this Protocol or under its domestic measures implementing the Protocol, may take into account, consistent with their international obligations, socio-economic considerations arising from the impact of living modified organisms on the conservation and sustainable use of biological diversity, especially with regard to the value of biological diversity to indigenous and local communities.
2. The Parties are encouraged to cooperate on research and information exchange on any socio-economic impacts of living modified organisms, especially on indigenous and local communities.

Article 27. Liability and Redress

The Conference of the Parties serving as the meeting of the Parties to this Protocol shall, at its first meeting, adopt a process with respect to the appropriate elaboration of international rules and procedures in the field of liability and redress for damage resulting from transboundary movements of living modified organisms, analysing and taking due account of the ongoing processes in international law on these matters, and shall endeavour to complete this process within four years.

Article 28. Financial Mechanism and Resources

1. In considering financial resources for the implementation of this Protocol, the Parties shall take into account the provisions of Article 20 of the Convention.
2. The financial mechanism established in Article 21 of the Convention shall, through the institutional structure entrusted with its operation, be the financial mechanism for this Protocol.
3. Regarding the capacity-building referred to in Article 22 of this Protocol, the Conference of the Parties serving as the meeting of the Parties to this Protocol, in providing guidance with respect to the financial mechanism referred to in paragraph 2 above, for consideration by the Conference of the Parties, shall take into account the need for financial resources by developing country Parties, in particular the least developed and the small island developing States among them.
4. In the context of paragraph 1 above, the Parties shall also take into account the needs of the developing country Parties, in particular the least developed and the small island developing States among them, and of the Parties with economies in transition, in their efforts to identify and implement their capacity-building requirements for the purposes of the implementation of this Protocol.
5. The guidance to the financial mechanism of the Convention in relevant decisions of the Conference of the Parties, including those agreed before the adoption of this Protocol, shall apply, *mutatis mutandis,* to the provisions of this Article.
6. The developed country Parties may also provide, and the developing country Parties and the Parties with economies in transition avail themselves of, financial and technological resources for the

implementation of the provisions of this Protocol through bilateral, regional and multilateral channels.

Article 29. Conference of the Parties Serving as the Meeting of the Parties to this Protocol

1. The Conference of the Parties shall serve as the meeting of the Parties to this Protocol.
2. Parties to the Convention that are not Parties to this Protocol may participate as observers in the proceedings of any meeting of the Conference of the Parties serving as the meeting of the Parties to this Protocol. When the Conference of the Parties serves as the meeting of the Parties to this Protocol, decisions under this Protocol shall be taken only by those that are Parties to it.
3. When the Conference of the Parties serves as the meeting of the Parties to this Protocol, any member of the bureau of the Conference of the Parties representing a Party to the Convention but, at that time, not a Party to this Protocol, shall be substituted by a member to be elected by and from among the Parties to this Protocol.
4. The Conference of the Parties serving as the meeting of the Parties to this Protocol shall keep under regular review the implementation of this Protocol and shall make, within its mandate, the decisions necessary to promote its effective implementation. It shall perform the functions assigned to it by this Protocol and shall:
   (a) Make recommendations on any matters necessary for the implementation of this Protocol;
   (b) Establish such subsidiary bodies as are deemed necessary for the implementation of this Protocol;
   (c) Seek and utilize, where appropriate, the services and cooperation of, and information provided by, competent international organizations and intergovernmental and non-governmental bodies;
   (d) Establish the form and the intervals for transmitting the information to be submitted in accordance with Article *33* of this Protocol and consider such information as well as reports submitted by any subsidiary body;

(e) Consider and adopt, as required, amendments to this Protocol and its annexes, as well as any additional annexes to this Protocol, that are deemed necessary for the implementation of this Protocol; and

(f) Exercise such other functions as may be required for the implementation of this Protocol.

5. The rules of procedure of the Conference of the Parties and financial rules of the Convention shall be applied, *mutatis mutandis,* under this Protocol, except as may be otherwise decided by consensus by the Conference of the Parties serving as the meeting of the Parties to this Protocol.

6. The first meeting of the Conference of the Parties serving as the meeting of the Parties to this Protocol shall be convened by the Secretariat in conjunction with the first meeting of the Conference of the Parties that is scheduled after the date of the entry into force of this Protocol. Subsequent ordinary meetings of the Conference of the Parties serving as the meeting of the Parties to this Protocol shall be held in conjunction with ordinary meetings of the Conference of the Parties, unless otherwise decided by the Conference of the Parties serving as the meeting of the Parties to this Protocol.

7. Extraordinary meetings of the Conference of the Parties serving as the meeting of the Parties to this Protocol shall be held at such other times as may be deemed necessary by the Conference of the Parties serving as the meeting of the Parties to this Protocol, or at the written request of any Party, provided that, within six months of the request being communicated to the Parties by the Secretariat, it is supported by at least one third of the Parties.

8. The United Nations, its specialized agencies and the International Atomic Energy Agency, as well as any State member thereof or observers thereto not party to the Convention, may be represented as observers at meetings of the Conference of the Parties serving as the meeting of the Parties to this Protocol. Any body or agency, whether national or international, governmental or non-governmental, that is qualified in matters covered by this Protocol and that has informed the Secretariat of its wish to be represented at a meeting of the Conference of the Parties serving as a meeting of the Parties to this Protocol as an

observer, may be so admitted, unless at least one third of the Parties present object. Except as otherwise provided in this Article, the admission and participation of observers shall be subject to the rules of procedure, as referred to in paragraph 5 above.

Article 30. Subsidiary Bodies

1. Any subsidiary body established by or under the Convention may, upon a decision by the Conference of the Parties serving as the meeting of the Parties to this Protocol, serve the Protocol, in which case the meeting of the Parties shall specify which functions that body shall exercise.
2. Parties to the Convention that are not Parties to this Protocol may participate as observers in the proceedings of any meeting of any such subsidiary bodies. When a subsidiary body of the Convention serves as a subsidiary body to this Protocol, decisions under the Protocol shall be taken only by the Parties to the Protocol.
3. When a subsidiary body of the Convention exercises its functions with regard to matters concerning this Protocol, any member of the bureau of that subsidiary body representing a Party to the Convention but, at that time, not a Party to the Protocol, shall be substituted by a member to be elected by and from among the Parties to the Protocol.

Article 31. Secretariat

1. The Secretariat established by Article 24 of the Convention shall serve as the secretariat to this Protocol.
2. Article 24, paragraph 1, of the Convention on the functions of the Secretariat shall apply, *mutatis mutandis,* to this Protocol.
3. To the extent that they are distinct, the costs of the secretariat services for this Protocol shall be met by the Parties hereto. The Conference of the Parties serving as the meeting of the Parties to this Protocol shall, at its first meeting, decide on the necessary budgetary arrangements to this end.

Article 32. Relationship With the Convention

Except as otherwise provided in this Protocol, the provisions of the Convention relating to its protocols shall apply to this Protocol.

Article 33. Monitoring and Reporting

Each Party shall monitor the implementation of its obligations under this Protocol, and shall, at intervals to be determined by the Conference of the Parties serving as the meeting of the Parties to this Protocol, report to the Conference of the Parties serving as the meeting of the Parties to this Protocol on measures that it has taken to implement the Protocol.

Article 34. Compliance

The Conference of the Parties serving as the meeting of the Parties to this Protocol shall, at its first meeting, consider and approve cooperative procedures and institutional mechanisms to promote compliance with the provisions of this Protocol and to address cases of non-compliance. These procedures and mechanisms shall include provisions to offer advice or assistance, where appropriate. They shall be separate from, and without prejudice to, the dispute settlement procedures and mechanisms established by Article 27 of the Convention.

Article 35. Assessment and Review

The Conference of the Parties serving as the meeting of the Parties to this Protocol shall undertake, five years after the entry into force of this Protocol and at least every five years thereafter, an evaluation of the effectiveness of the Protocol, including an assessment of its procedures and annexes.

Article 36. Signature

This Protocol shall be open for signature at the United Nations Office at Nairobi by States and regional economic integration organizations from 15 to 26 May 2000, and at United Nations Headquarters in New York from 5 June 2000 to 4 June 2001.

Article 37. Entry Into Force

1. This Protocol shall enter into force on the ninetieth day after the date of deposit of the fiftieth instrument of ratification, acceptance, approval or accession by States or regional economic integration organizations that are Parties to the Convention.

2. This Protocol shall enter into force for a State or regional economic integration organization that ratifies, accepts or approves this Protocol or accedes thereto after its entry into force pursuant to paragraph 1 above, on the ninetieth day after the date on which that State or regional economic integration organization deposits its instrument of ratification, acceptance, approval or accession, or on the date on which the Convention enters into force for that State or regional economic integration organization, whichever shall be the later.
3. For the purposes of paragraphs 1 and 2 above, any instrument deposited by a regional economic integration organization shall not be counted as additional to those deposited by member States of such organization.

Article 38. Reservations

No reservations may be made to this Protocol.

Article 39. Withdrawal

1. At any time after two years from the date on which this Protocol has entered into force for a Party, that Party may withdraw from the Protocol by giving written notification to the Depositary.
2. Any such withdrawal shall take place upon expiry of one year after the date of its receipt by the Depositary, or on such later date as may be specified in the notification of the withdrawal.

Article 40. Authentic Texts

The original of this Protocol, of which the Arabic, Chinese, English, French, Russian and Spanish texts are equally authentic, shall be deposited with the Secretary-General of the United Nations.

*In Witness Whereof*the undersigned, being duly authorized to that effect, have signed this Protocol.

Done at Montréal on this twenty-ninth day of January, two thousand.

Annex I. Information Required In Notifications Under Articles 8,10 And 13

(a) Name, address and contact details of the exporter.

(b) Name, address and contact details of the importer.

(c) Name and identity of the living modified organism, as well as the domestic classification, if any, of the biosafety level of the living modified organism in the State of export.

(d) Intended date or dates of the transboundary movement, if known.

(e) Taxonomic status, common name, point of collection or acquisition, and characteristics of recipient organism or parental organisms related to biosafety.

(f) Centres of origin and centres of genetic diversity, if known, of the recipient organism and/or the parental organisms and a description of the habitats where the organisms may persist or proliferate.

(g) Taxonomic status, common name, point of collection or acquisition, and characteristics of the donor organism or organisms related to biosafety.

(h) Description of the nucleic acid or the modification introduced, the technique used, and the resulting characteristics of the living modified organism.

(i) Intended use of the living modified organism or products thereof, namely, processed materials that are of living modified organism origin, containing detectable novel combinations of replicable genetic material obtained through the use of modern biotechnology.

(j) Quantity or volume of the living modified organism to be transferred.

(k) A previous and existing risk assessment report consistent with Annex III.

(l) Suggested methods for the safe handling, storage, transport and use, including packaging, labelling, documentation, disposal and contingency procedures, where appropriate.

(m) Regulatory status of the living modified organism within the State of export (for example, whether it is prohibited in the State of export, whether there are other restrictions, or whether it has been approved for general release) and, if the living modified organism is banned in the State of export, the reason or reasons for the ban.

(n) Result and purpose of any notification by the exporter to other States regarding the living modified organism to be transferred.

(o) A declaration that the above-mentioned information is factually correct.

Annex II. Information Required Concerning Living Modified Organisms Intended for Direct Use as Food or Feed, or for Processing Under Article 11

(a) The name and contact details of the applicant for a decision for domestic use.

(b) The name and contact details of the authority responsible for the decision.

(c) Name and identity of the living modified organism.

(d) Description of the gene modification, the technique used, and the resulting characteristics of the living modified organism.

(e) Any unique identification of the living modified organism.

(f) Taxonomic status, common name, point of collection or acquisition, and characteristics of recipient organism or parental organisms related to biosafety.

(g) Centres of origin and centres of genetic diversity, if known, of the recipient organism and/or the parental organisms and a description of the habitats where the organisms may persist or proliferate.

(h) Taxonomic status, common name, point of collection or acquisition, and characteristics of the donor organism or organisms related to biosafety.

(i) Approved uses of the living modified organism.

(j) A risk assessment report consistent with Annex III.

(k) Suggested methods for the safe handling, storage, transport and use, including packaging, labelling, documentation, disposal and contingency procedures, where appropriate.

Annex III. Risk Assessment

Objective

1. The objective of risk assessment, under this Protocol, is to identify and evaluate the potential adverse effects of living modified organisms on the conservation and sustainable use of biological diversity in the likely potential receiving environment, taking also into account risks to human health.

Use of Risk Assessment

2. Risk assessment is, *inter alia,* used by competent authorities to make informed decisions regarding living modified organisms.

General Principles

3. Risk assessment should be carried out in a scientifically sound and transparent manner, and can take into account expert advice of, and guidelines developed by, relevant international organizations.
4. Lack of scientific knowledge or scientific consensus should not necessarily be interpreted as indicating a particular level of risk, an absence of risk, or an acceptable risk.
5. Risks associated with living modified organisms or products thereof, namely, processed materials that are of living modified organism origin, containing detectable novel combinations of replicable genetic material obtained through the use of modern biotechnology, should be considered in the context of the risks posed by the non-modified recipients or parental organisms in the likely potential receiving environment.
6. Risk assessment should be carried out on a case-by-case basis. The required information may vary in nature and level of detail from case to case, depending on the living modified organism concerned, its intended use and the likely potential receiving environment.

Methodology

7. The process of risk assessment may on the one hand give rise to a need for further information about specific subjects, which may be identified and requested during the assessment process, while on the other hand information on other subjects may not be relevant in some instances.
8. To fulfil its objective, risk assessment entails, as appropriate, the following steps:
   (a) An identification of any novel genotypic and phenotypic characteristics associated with the living modified organism that may have adverse effects on biological diversity in the likely potential receiving environment, taking also into account risks to human health;

(b) An evaluation of the likelihood of these adverse effects being realized, taking into account the level and kind of exposure of the likely potential receiving environment to the living modified organism;

(c) An evaluation of the consequences should these adverse effects be realized;

(d) An estimation of the overall risk posed by the living modified organism based on the evaluation of the likelihood and consequences of the identified adverse effects being realized;

(e) A recommendation as to whether or not the risks are acceptable or manageable, including, where necessary, identification of strategies to manage these risks; and

(f) Where there is uncertainty regarding the level of risk, it may be addressed by requesting further information on the specific issues of concern or by implementing appropriate risk management strategies and/or monitoring the living modified organism in the receiving environment.

Points to Consider

9. Depending on the case, risk assessment takes into account the relevant technical and scientific details regarding the characteristics of the following subjects:

(a) *Recipient organism or parental organisms.* The biological characteristics of the recipient organism or parental organisms, including information on taxonomic status, common name, origin, centres of origin and centres of genetic diversity, if known, and a description of the habitat where the organisms may persist or proliferate;

(b) *Donor organism or organisms.* Taxonomic status and common name, source, and the relevant biological characteristics of the donor organisms;

(c) *Vector.* Characteristics of the vector, including its identity, if any, and its source or origin, and its host range;

(d) *Insert or inserts and/or characteristics of modification.* Genetic characteristics of the inserted nucleic acid and the function it specifies, and/or characteristics of the modification introduced;

(e) *Living modified organism.* Identity of the living modified organism, and the differences between the biological characteristics of the living modified organism and those of the recipient organism or parental organisms;

(f) *Detection and identification of the living modified organism.* Suggested detection and identification methods and their specificity, sensitivity and reliability;

(g) *Information relating to the intended use.* Information relating to the intended use of the living modified organism, including new or changed use compared to the recipient organism or parental organisms; and

(h) *Receiving environment.* Information on the location, geographical, climatic and ecological characteristics, including relevant information on biological diversity and centres of origin of the likely potential receiving environment.

Bibliography

Barbault, R. and S. D. Sastrapradja. (1995). *Generation, maintenance and loss of biodiversity. Global Biodiversity Assessment,* Cambridge Univ. Press, Cambridge pp. 193–274.

Bowen B. W. (1999). "Preserving genes, species, or ecosystems? Healing the fractured foundations of conservation policy"(PDF). *Molecular Ecology* 8: S5–S10.

Brooks T. M., Mittermeier R. A., Gerlach J., Hoffmann M., Lamoreux J. F., Mittermeier C. G., Pilgrim J. D., Rodrigues A. S. L. (2006). "Global Biodiversity Conservation Priorities". *Science* 313 (5783): 58.

Canadell, J.G.; M.R. Raupach (2008-06-13). "Managing Forests for Climate Change". *Science*(AAAS) 320 (5882): 1456–1457.

Clarke A. & Harris C.M., (2003)."Polar marine ecosystems: major threats and future change", *Environmental Conservation.*

Cooper, Chris; et al (2005). *Tourism: Principles and Practice* (3rd ed.). Harlow: Pearson Education.

Dudley, Nigel (1988); *Acid Rain and Wildlife*: A Report to Wildlife Link, Earth Resources Research, London.

Harter, P (1988); *Acid Deposition: Ecological Effects*, IEA Coal Research, London.

Hickling R., Roy D.B., Hill J.K., Fox R. & Thomas C.D., (2006)."The distributions of a wide range of taxonomic groups are expanding northwards", *Global Change Biology.*

Houghton J., (2002). *Global Warming the Complete Briefing*, Cambridge University Press,

Isaacs, J.C. (2000). *The limited potential of ecotourism to contribute to wildlife conservation*. The Ecologist. pp. 28(1):61–69.

Kareiva P., Marvier M. (2003). "Conserving Biodiversity Coldspots" (PDF). *American Scientist* 91 (4): 344–351.

Leveque, C. & J. Mounolou (2003) *Biodiversity*. New York: John Wiley.

Margulis, L., Dolan, Delisle, K., Lyons, C. *Diversity of Life: The Illustrated Guide to the Five Kingdoms*. Sudbury: Jones & Bartlett Publishers.

McCallum M. L. (2008). "Amphibian Decline or Extinction? Current Declines Dwarf Background Extinction Rate" (PDF).*Journal of Herpetology* 41 (3): 483–491.

McLaren, D. (1998). *Rethinking tourism and ecotravel: the paving of paradise and what you can do to stop it.* West Hartford, Connecticut, USA: Kamarian Press.

MEA. (2005). *Ecosystems and Human Well-Being. Millennium Ecosystem Assessment.* Island Press, Covelo, CA.

Myers N., Mittermeier R. A., Mittermeier C. G., Kent J. (2000). "Biodiversity hotspots for conservation priorities". *Nature* 403: 853–858.

Novacek, M. J. (ed.) (2001) *The Biodiversity Crisis: Losing What Counts.* New York: American Museum of Natural HistoryBooks.

Pereira, H. M.; Navarro, L. M.; Martins, I. S. S. (2012). "Global Biodiversity Change: The Bad, the Good, and the Unknown".*Annual Review of Environment and Resources* 37: 25.

Primack, Richard B. (2004). *A primer of Conservation Biology.* Sunderland, Mass: Sinauer Associates.

Singh, L. K. (2008). Issues in Tourism Industry. *Fundamental of Tourism and Travel.* Delhi: Isha Books.

Stein, B. A., L. S. Kutner, and J. S. Adams (eds.). (2000). *Precious Heritage: The Status of Biodiversity in the United States.* Oxford University Press, New York.

Tickle, Andrew, with Malcolm Fergusson and Graham Drucker; Acid Rain and Nature *Conservation in Europe: A preliminary study of areas at risk from acidification,* WWF International, Gland, Switzerland.

Ulrich, B and J Pankrath [editors] (1983) *Effects of Accumulation of Air Pollutants in Forest Ecosystems,* Reidel Publishers.

Virtanen, T. and S. Neuvonen, (1999). "Climate change and macrolepidopteran biodiversity in Finland", *Chemosphere: Global Change Science.*

Vivanco, L. (2002). *Ecotourism, Paradise lost—A Thai case study.* The Ecologist. pp. 32(2):28–30.

Wallisdevries M.F. & Van Swaay C.A.M., (2006)."Global warming and excess nitrogen may induce butterfly decline by microclimatic cooling", *Global Change Biology.*

White, R. P., S. Murray, and M. Rohweder. (2000). *Pilot Assessment of Global Ecosystems: Grassland Ecosystems.* World Resources Institute, Washington, DC.

Whitney, Gordon G. (1996). *From Coastal Wilderness to Fruited Plain : A History of Environmental Change in Temperate North America from 1500 to the Present.* Cambridge University Press.

Williams, Michael. (2003). *Deforesting the Earth.* University of Chicago Press, Chicago.

Wood well, G.M.; North, WJ (1988-12-16). "CO2Reduction and Reforestation". *Science* (ALAS) 242(4885): 1493–1494.

Wunder, Sven. (2000). *The Economics of Deforestation: The Example of Ecuador.* Macmillan Press, London.